Student Solutions Manual

Miller & Freund's

Probability and Statistics for Engineers

SEVENTH EDITION

Richard A. Johnson

University of Wisconsin, Madison

Upper Saddle River, New Jersey 07458

Editor-in-Chief: Sally Yagan
Acquisitions Editor: George Lobell
Supplements Editor: Jennifer Urban
Vice President of Production & Manufacturing: David W. Riccardi
Executive Managing Editor: Kathleen Schiaparelli
Managing Editor: Becca Richter
Production Editor: Donna Crilly
Manufacturing Manager: Trudy Pisciotti
Assistant Manufacturing Manager: Michael Bell
Manufacturing Buyer: Ilene Kahn
Supplement Cover Designer: Elizabeth Wright

Pearson Prentice Hall
Pearson Education, Inc.
Upper Saddle River, NJ 07458

Printed in the United States of America

10 9 8 7 6 5 4 3

ISBN 0-13-143746-1

Pearson Education Ltd., *London*
Pearson Education Australia Pty. Ltd., *Sydney*
Pearson Education Singapore, Pte. Ltd.
Pearson Education North Asia Ltd., *Hong Kong*
Pearson Education Canada, Inc., *Toronto*
Pearson Educación de Mexico, S.A. de C.V.
Pearson Education—Japan, *Tokyo*
Pearson Education Malaysia, Pte. Ltd.

Contents

PREFACE

This students' manual is intended to help the student gain an improved understanding of the subject by providing model solutions for all of the odd numbered exercises in the text. The chapters in this manual correspond to those in the text.

In the spirit of quality improvement, we would appreciate receiving your comments, corrections and suggestions for improvements.

Richard A. Johnson

Chapter 1

INTRODUCTION

1.1 The statistical *population* could be the collection of air quality values for all U. S. based flights flown during the period of the study. It could also be expanded to include all flights for the year or even all those that could have conceivably been flown. The *sample* consists of the measurements from the 158 actual flights on which air quality was measured.

1.3 a) A laptop owned by student.

b) Size of hard disk.

c) Collection of numbers, for all student owned laptops, specifying size of the hard disk.

1.5 We used the first page of Table 7, row 11, and columns 17 and 18. Reading down, we ignore 93, 80 and select 31, 4, and 29. After ignoring several other numbers greater than 40, we get 21 and 24. We select the five persons at these positions on the active membership role.

1.7 a) For the new sample, $\overline{x} = 214.67$ and the X-bar chart is shown in Figure 1.1

b) The new $\overline{x}$ is below the lower control limit LCL $= 215$ so the process is not yet stable.

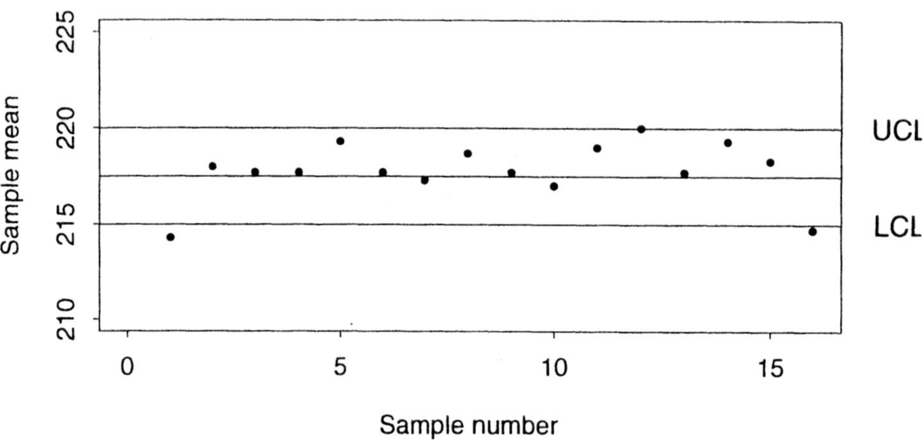

Figure 1.1: X-bar chart for slot depth. Exercise 1.7

Chapter 2

TREATMENT OF DATA

2.1 Pareto chart of the accident data

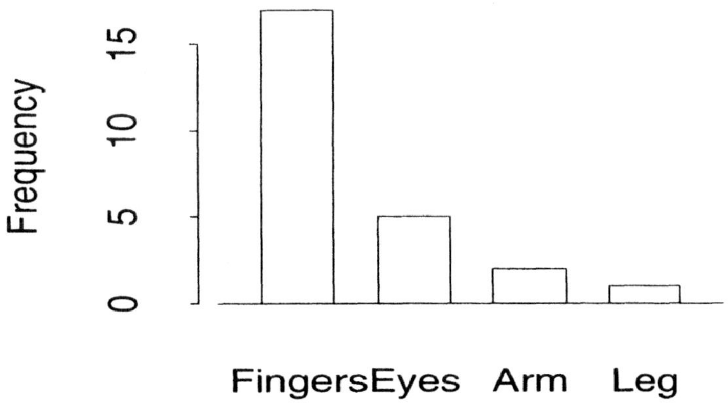

Figure 2.1: Pareto chart for Exercise 2.1

2.3 The dot diagram of the boiling point of a silicon compound data.

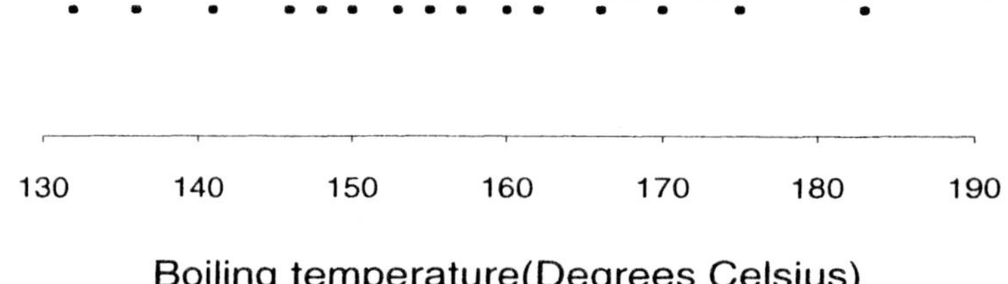

2.5 The dot diagram of the suspended solids data reveals that one reading, 65 ppm, is very large. Other readings taken at about the same time, but not given here, confirm that the water quality was bad at that time. That is, 65 is a reliable number for that day.

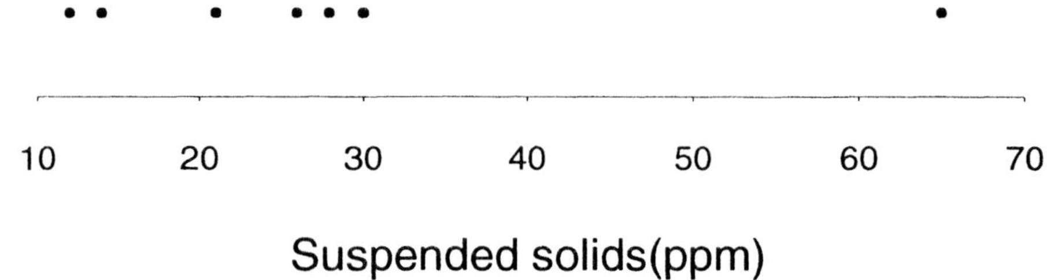

2.7 a) The class marks are 11.0, 12.0, 13.0, and 14.0

b) The class interval is 1.

2.9 Grouping the data gives the following table:

Class limits	Frequency
[15, 20)	3
[20, 25)	15
[25, 30)	24
[30, 35)	12
[35, 40)	6

The histogram is given below.

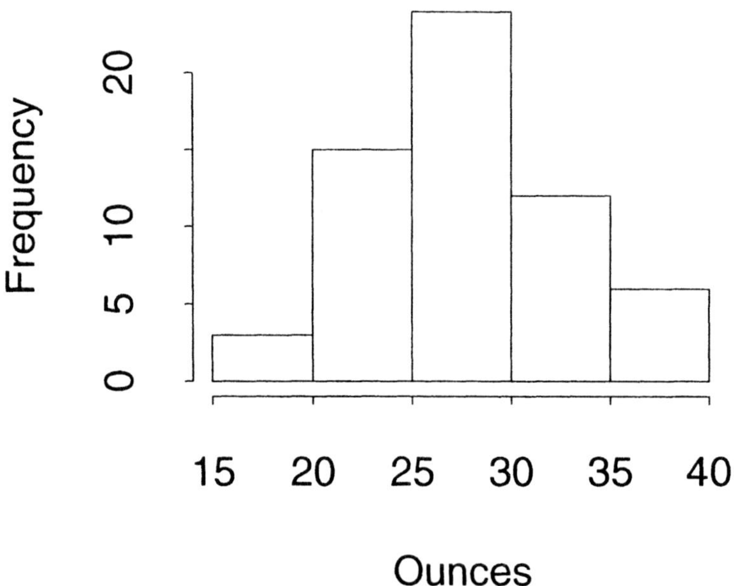

2.11 a) The class boundaries are symmetric around the class marks, and all but the two most extreme are given by the mid-points between the class marks. Thus, 11.5 , 20.5 , 29.5, 38.5, 47.5, 56.5, 65.5 determine the boundaries of the classes.

b) The class interval is 9.

2.13 We first convert the data in the preceding exercise to a "less than" distribution. The ogive is plotted below in Figure 2.2.

x	No. less than	x	No. less than
1.0	0	8.0	68
2.0	10	9.0	72
3.0	20	10.0	76
4.0	29	11.0	77
5.0	40	12.0	79
6.0	52	13.0	80
7.0	62		

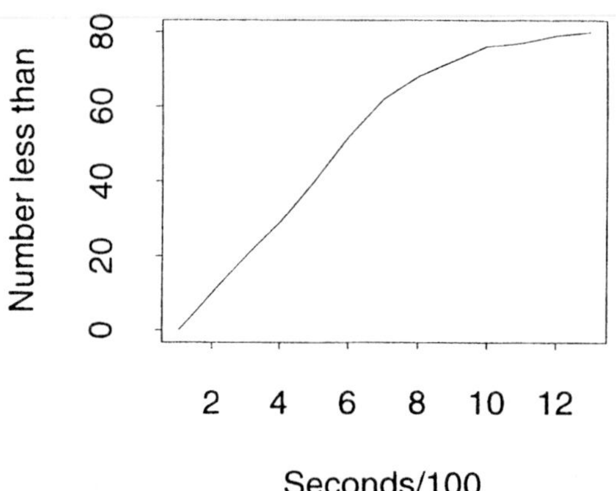

Figure 2.2: Ogive for Exercise 2.13

2.15 The "less than" distribution of the data in the preceding exercise is:

Class boundary	Number less than	Class boundary	Number less than
20.0	0	60.0	60
30.0	4	70.0	80
40.0	17	80.0	94
50.0	35	90.0	100

The ogive is plotted in Figure 2.3.

2.17 The cumulative "or more" distribution is:

Class boundary	Number equal or more	Class boundary	Number equal or more
-0.5	60	3.5	15
0.5	45	4.5	7
1.5	33	5.5	2
2.5	22	6.5	0

The ogive is shown in Figure 2.4.

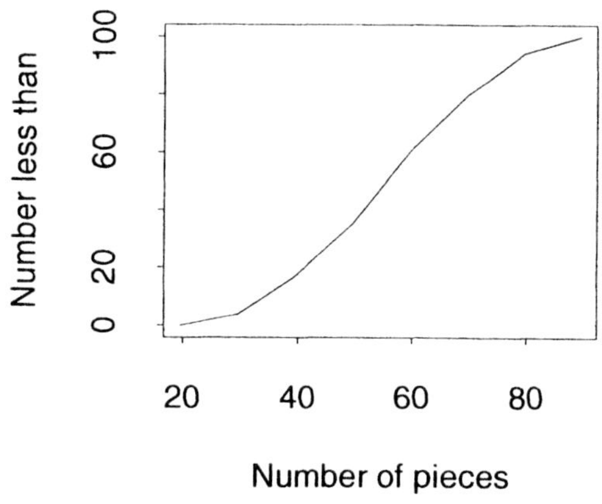

Figure 2.3: Ogive for Exercise 2.15

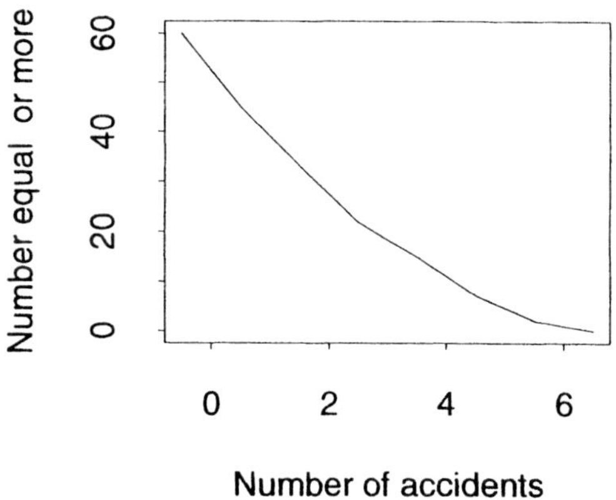

Figure 2.4: Ogive for Exercise 2.17

2.19 No. We tend to compare areas visually. The area of the large sack is far more than double the area of the small sack. The large sack should be modified so that its area is double that of the small sack.

2.21 The empirical cumulative distribution function is

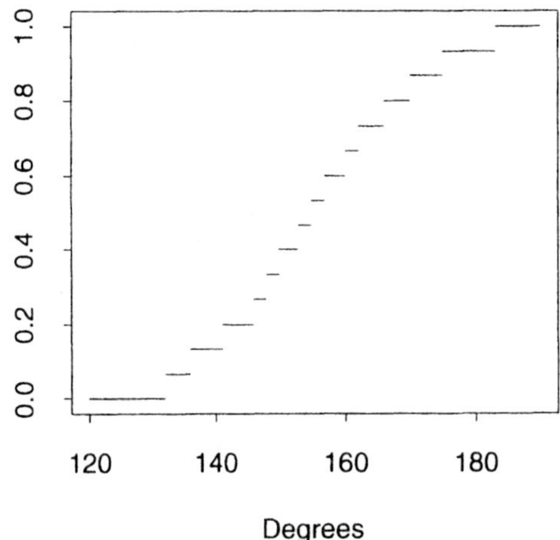

2.23 The stem and leaf display is:

1**	
2**	67, 88, 95
3**	55, 70, 91, 83, 17
4**	05, 19, 34, 62
5**	40, 08
6**	12
7**	

2.25 The stem-and-leaf display is:

2*	2 1
2·	8 6
3*	2 4 3 4
3·	6 6 7 9 5 5 5 5 8
4*	3 3 0 0 4 2 1 1
4·	9 6 8 5 8 5 5 5 7 7
5*	0 1 2 3 1 4 1 3 2 2 0 3 4 0
5·	6 5 6 7 7 9 6 5 5 9 8
6*	2 0 2 3 4 0 1 1 0 2 1
6·	5 7 8 5 8 7 9 8 5
7*	2 3 4 0 4 3 4 0
7·	9 8 6 5 6 7
8*	4 2 0 2
8·	8 5

2.27 (a) Would like a high salary-outlier that is large.

(b) Would like a high score-outlier that is large.

(c) Near average.

2.29 Greater on the mean. It would not influence the median for sample size 3 or more.

2.31 (a) $\overline{x} = (-6 + 1 - 4 - 3)/4 = -3$

(b)

$$s^2 = \frac{(-6-(-3))^2 \;+\; (1-(-3))^2 \;+\; (-4-(-3))^2 \;+\; (-3-(-3))^2}{4-1}$$

$$= (9 + 16 + 1 + 0)/3 = 8.667 \quad \text{so} \quad s = \sqrt{8.667} = 2.9.$$

(c) The mean of the observation minus specification is -3 so, on average, the observations are below the specification. On average, the holes are too small.

2.33 For the non-leakers

$$\overline{x} = (.207 + .124 + .062 + .301 + .186 + .124)/6 = \frac{1.004}{6} = .16733$$

and $.207^2 + .124^2 + .062^2 + .301^2 + .186^2 + .124^2 = .20264$ so

$$s^2 = \frac{6(.20264) - (1.004)^2}{6 \cdot 5} = .006928$$

Consequently, $s = \sqrt{.006928} = .0832$. The means are quite close to each other and so are the standard deviations. The size of gap does not seem to be connected to the existence of leaks.

2.35 No. The sum of the salaries must be equal to 3 (125,000) = 375,000 which is less than 400,000. This assumes that all salaries are non-negative. It is certainly possible if negative salaries are allowed.

2.37 (a) The mean is:

$$\frac{33 + 24 + 39 + 48 + 26 + 35 + 38 + 54 + 23 + 34 + 29 + 37}{12} = 35$$

(b) Sorting the data gives:

23, 24, 26, 29, 33, 34, 35, 37, 38, 39, 48, 54.

The median is the average of the sixth and seventh smallest observations or $(34 + 35)/2 = 34.5$.

2.39 (a) The mean is 8.

(b) The sorted data are:

1, 2, 2, 3, 5, 6, 8, 9, 9, 10, 10, 10, 13, 15, 17.

The median is the eighth smallest which is 9.

(c) The boxplot is given in Figure 2.5.

2.41 The mean is 30.91. The sorted data are:

29.6, 30.3, 30.4, 30.5, 30.7, 31.0, 31.2, 31.2, 32.0, 32.2

Since the number of observations is 10 (an even number), the median is the average

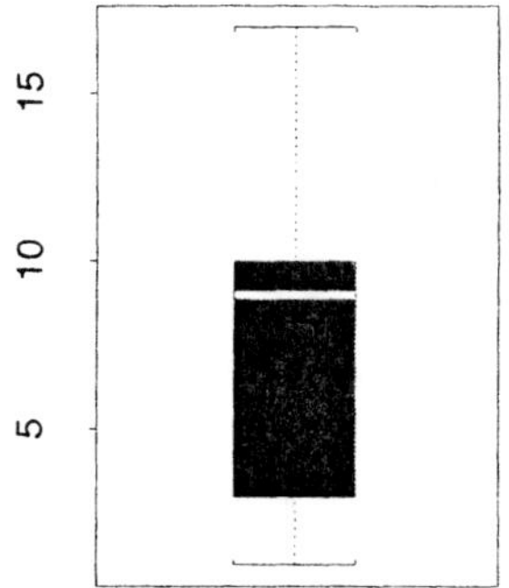

Figure 2.5: Boxplot for Exercise 2.39

of the 5'th and 6'th observations or $(30.7 + 31.0)/2 = 30.85$. The first quartile is the third observation, 30.4, and the third quartile is the eighth observation, 31.2.

2.43 The box plot is given in Figure 2.6.

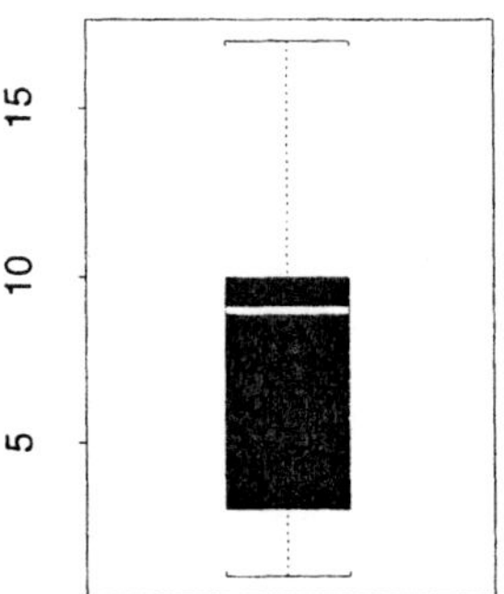

Figure 2.6: Box-plot for Exercise 2.43

2.45 (a) $\sum x_i = 605$. Thus, $\bar{x} = 605/20 = 30.25$, and $\sum x_i^2 = 20,663$. Hence,

$s^2 = (20 \cdot 21723 - 615^2)/(20 \cdot 19) = 124.303$. And $s = 11.15$.

(b) The class limits, marks, and frequencies are in the following table:

Class limits	Class mark	Frequency
10 - 19	14.5	3
20 - 29	24.5	8
30 - 39	34.5	5
40 - 49	44.5	3
50 - 59	54.5	1

Thus,

$\bar{x} = (3(14.5) + 8(24.5) + 5(34.5) + 3(44.5) + 54.5)/20 = 30.$

$\sum x_i f_i = 600$. $\sum x_i^2 f_i = 20,295$. Thus ,

$s^2 = (20 \cdot 20,295 - 600^2)/(20 \cdot 19) = 120.79$. So, $s = 10.99$.

2.47 We use the class limits [1, 2), [2,3) , ... where the righthand endpoint is excluded. The class marks and frequencies are:

Class mark	Frequency
1.5	10
2.5	10
3.5	9
4.5	11
5.5	12
6.5	10
7.5	6
8.5	4
9.5	4
10.5	1
11.5	2
12.5	1

Thus, $\sum x_i f_i = 415.0$, $\sum x_i^2 f_i = 2,722.0$. Therefore, the mean is $415.0/80 = 5.188$ and $s^2 = (80 \cdot 2,722.0 - 415.0^2)/(80 \cdot 79) = 7.2049$. The standard deviation is $s = 2.684$. The coefficient of variation is $v = 100\, s/\bar{x} = 51.0$ percent.

2.49 Let x_i be the copper prices, and y_i be the coal prices. then, $\bar{x} = 67.57$ and $s_{copper} = 1.779$. Thus, $v_{copper} = 2.63$ percent. Now $\bar{y} = 20.55$ and $s_{coal} = 1.614$. Thus, $v_{coal} = 7.85$ percent. Thus, coal prices are relatively more variable.

2.51

$$\left(\sum_{i=1}^{n}(x_i - \bar{x})^2\right)/(n-1) = \sum_{i=1}^{n}(x_i^2 - 2x_i\bar{x} + \bar{x}^2)/(n-1)$$

$$= \left(\sum_{i=1}^{n} x_i^2\right)/(n-1) - \left(2\bar{x}\sum_{i=1}^{n} x_i\right)/(n-1) + n\bar{x}^2/(n-1)$$

$$= \left(\sum_{i=1}^{n} x_i^2\right)/(n-1) - 2\left(\sum_{i=1}^{n} x_i\right)^2/(n(n-1)) + \left(\sum_{i=1}^{n} x_i\right)^2/(n(n-1))$$

$$= \left(\sum_{i=1}^{n} x_i^2\right)/(n-1) - \left(\sum_{i=1}^{n} x_i\right)^2/(n(n-1)) = \left(n\sum_{i=1}^{n} x_i^2 - \left(\sum_{i=1}^{n} x_i\right)^2\right)/(n(n-1))$$

2.53 (a) The median is the average of the 25'th and 26'th largest observations. These values are in the third class (10 - 14) which has frequency 16. The lower class boundary is 9.5 and the class interval is 5. There are 19 observations in the lower two classes. Thus, the estimate for the median is $5(50/2 - 19)/16 + 9.5 = 11.375$.

(b) There are 60 observations. The median falls in the third class (25.0 - 29.9) which has frequency 24. There are 18 observations in the lower two classes. The lower class boundary is 24.95 and the class interval is 5. Thus, the estimate for the median is $5(60/2 - 18)/24 + 24.95 = 27.45$.

(c) There are 80 observations. There are 40 observations in the first four classes, and 40 in the last four. Thus, the estimate for the median is the class boundary between the 4'th and 5'th class or 4.995.

2.55 (a) There are 80 observations. Q_1 is in the third class and Q_3 is in the 5'th class. The lower class boundary for the 3'rd class is 12.95. There are 13 observations in the first two classes and 14 in the third class. The class interval is 4. Thus,

Q_1 is estimated by $4(80/4 - 13)/14 + 12.95 = 14.95$.

Proceeding in a similar fashion gives $20.95 + 4(60 - 52)/17 = 22.83$ as the estimate of Q_3. Thus, the interquartile range is 7.88.

(b) the estimate for Q_1 is $5(50/4 - 4)/15 + 4.5 = 7.33$, and for Q_3 it is $5((3/4)50 - 35)/8 + 14.5 = 16.06$.

2.57 (a) The weighted average for the student is

$(69 + 75 + 56 + 72 + 4 \cdot 78)/8 = 73.0$.

(b) The combined percent increase for the average salaried worker is:

$(28 \cdot 53 + 35 \cdot 40 + 14 \cdot 34)/(28 + 35 + 14) = 43.64$ percent.

2.59 (a) From the ordered data, the first quartile $Q_1 = 1,712$, the median $Q_2 = 1,863$ and the third quartile $Q_3 = 2,061$.

(b) The histogram is given in Figure 2.7.

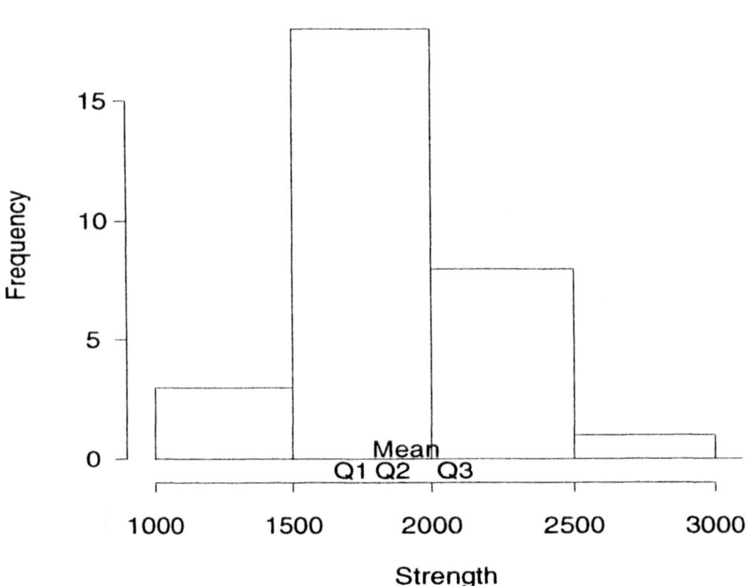

Figure 2.7: Exercise 2.59(b): Mean = 1,909, $Q_1 = 1,712$, $Q_2 = 1,863$, $Q_3 = 2,061$.

(c) For the aluminum data, we first sort the data.

66.4	67.7	68.0	68.0	68.3	68.4	68.6	68.8
68.9	69.0	69.1	69.2	69.3	69.3	69.5	69.5
69.6	69.7	69.8	69.8	69.9	70.0	70.0	70.1
70.2	70.3	70.3	70.4	70.5	70.6	70.6	70.8
70.9	71.0	71.1	71.2	71.3	71.3	71.5	71.6
71.6	71.7	71.8	71.8	71.9	72.1	72.2	72.3
72.4	72.6	72.7	72.9	73.1	73.3	73.5	74.2
74.5	75.3						

The first quartile $Q_1 = 69.5$, the median $Q_2 = 70.55$ and the third quartile $Q_3 = 71.80$. The histogram is given in Figure 2.8.

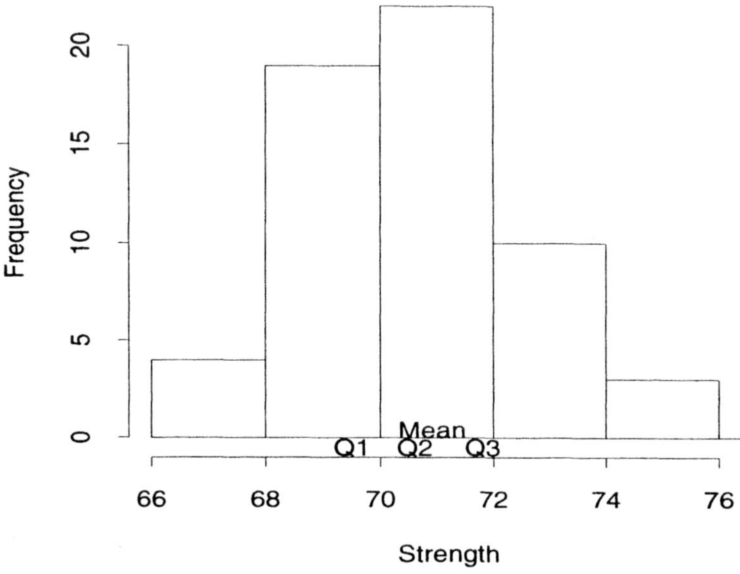

Figure 2.8: Exercise 2.59(c): Mean $= 70.9$, $Q_1 = 69.5$, $Q_2 = 70.55$, $Q_3 = 71.80$

2.61 (a) The frequency table of the aluminum alloy strength data (righthand endpoint excluded) is

Class limits	Frequency
$[66.0, 67.5)$	1
$[67.5, 69.0)$	8
$[69.0, 70.5)$	19
$[70.5, 72.0)$	17
$[72.0, 73.5)$	9
$[73.5, 75.9)$	3
$[75.0, 76.5)$	1

(b) The histogram, using the frequency table in part (a), is shown in Figure 2.9.

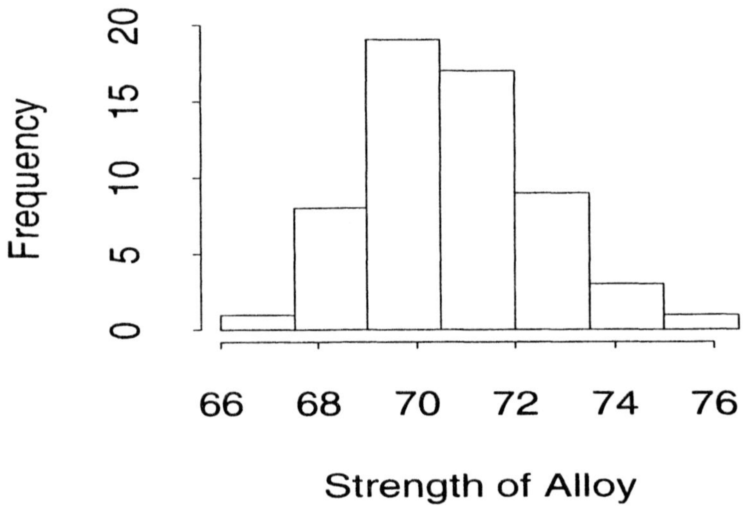

Figure 2.9: Histogram for Exercise 2.61

2.63 (a) The mean and standard deviation for the earth's density data are

$\bar{x} = 5.4835$ and $s = 0.19042$

(b) The ordered data are

5.10, 5.27, 5.29, 5.29, 5.30, 5.34, 5.34, 5.36, 5.39, 5.42, 5.44, 5.46,

5.47, 5.53, 5.57, 5.58, 5.62, 5.63, 5.65, 5.68, 5.75, 5.79, 5.85

There are $n = 23$ observations. The median is the middle value, or 5.46. There must be at least 23/4=5.75 observations at or below Q_1, and at least 23(3/4)=17.25 observations at or above Q_1. Only $Q_1 = 5.34$ satisfies this condition. Similarly, there must be at least 3(23/4)=17.25 observations at or below Q_3, and at least 23/4=5.75 observations at or above Q_3. Only $Q_3 = 5.63$ satisfies this condition.

(c) From the plot of the observations versus time order we see that there is no obvious trend although there is some suggestion of an increase over the last half of the observations.(See Figure 2.10)

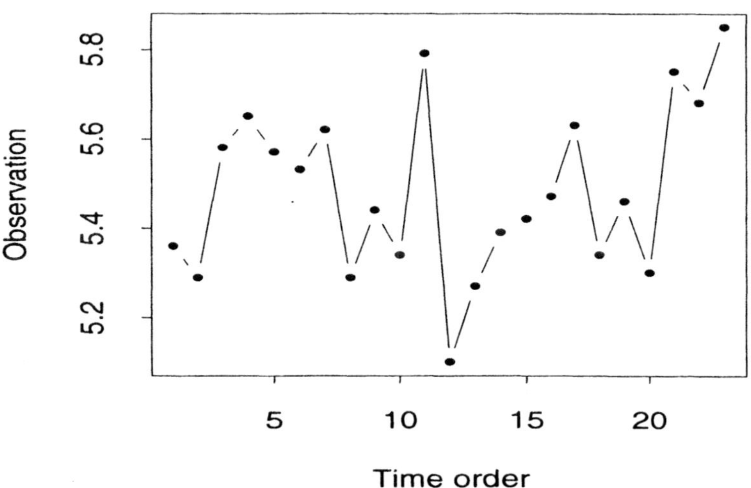

Figure 2.10: Plot of the observations vs time order, Exercise 2.63

2.65 (a) The ordered data are:

0.32 0.34 0.40 0.40 0.43 0.48 0.57

Since there are 7 observations, the median is the middle value. The median, maximum, minimum and range for the Tube 1 observations are:

Median = 0.40, maximum = 0.57, minimum = 0.32 and

range = maximum − minimum = 0.57 − 0.32 = 0.25.

(b) The ordered data are:

0.47 0.47 0.48 0.51 0.53 0.61 0.63

And the median, maximum, minimum and range for the Tube 2 observations are:

Median = 0.51, maximum = 0.63, minimum = 0.47 and

range = maximum − minimum = 0.63 − 0.47 = 0.16.

2.67 (a) The quartiles for the velocity of light data are

$Q_1 = 18.0$ and $Q_3 = 30.0$

(b) The minimum, maximum, range and the interquartile range are

Minimum = 12, maximum = 48,

range = maximum − minimum = 48 − 12 = 36 and

interquartile range $= Q_3 - Q_1 = 30.0 - 18.0 = 12.0$.

(c) The box-plot is shown in Figure 2.11.

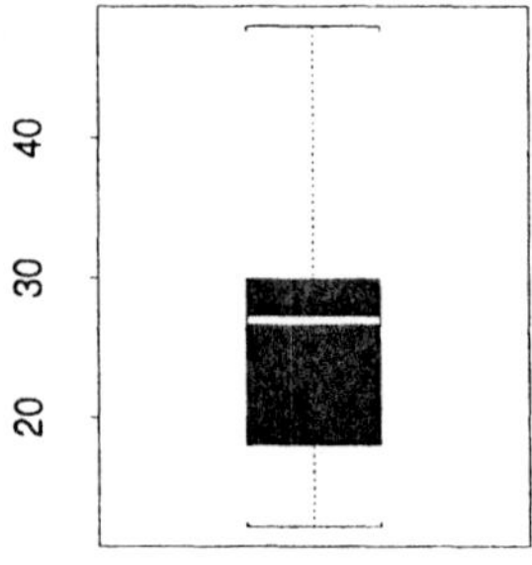

Figure 2.11: Box-Plot for Exercise 2.67

2.69 (a) The ordered data are:

12 14 19 20 21 28 29 30 55 63 63

The quartiles for the suspended solids data are $Q_1 = 19$ and $Q_3 = 55$.

(b) The minimum, maximum, range and the interquartile range are

Minimum = 12, maximum = 63, range = 63 − 12 = 51 and

interquartile range = $Q_3 - Q_1$ = 55 − 19 = 36.

(c) The boxplot is

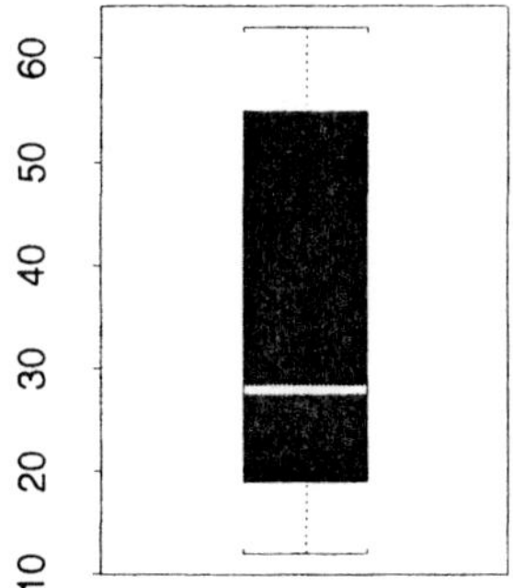

2.71 Boxplot of the aluminum data is shown in Figure 2.12

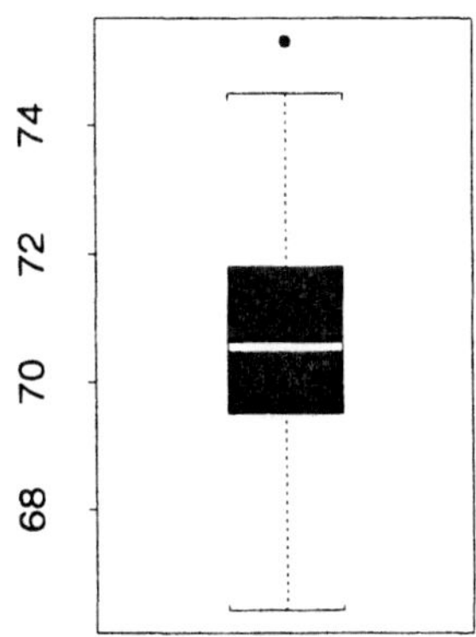

Figure 2.12: Boxplot for Exercise 2.71

2.73 The coefficient of variation for the first student is

$$\frac{3.3}{63.2} \cdot 100 = 5.22 \text{ percent.}$$

The coefficient of variation for the second student is

$$\frac{5.3}{78.8} \cdot 100 = 6.726 \text{ percent.}$$

Thus, the first student is relatively more consistent.

2.75 (a) The ordered observations are

389.1 390.8 392.4 400.1 425.9 429.1 448.4 461.6
479.1 480.8 482.9 497.2 505.8 516.5 517.5 547.5
550.9 563.7 567.7 572.2 572.5 575.6 595.5 602.0
606.7 611.9 618.9 626.9 634.9 644.0 657.6 679.3
698.6 718.5 738.0 743.3 752.6 760.6 794.8 817.2
833.9 889.0 895.8 904.7 986.4 1146.0 1156.0

The first quartile is the 12th observation, 497.2, the median is the 24th observation, 602.0, and the third quartile is the 36th observation, 743.3.

(b) Since 47(.90) = 42.3, the 90th percentile is the 43rd observation, 895.8.

(c) The histogram is shown in Figure 2.13

2.77 No. To be an outlier, the minimum or maximum must be quite separated from the next few closest observations.

2.79 (a) The dot diagram in Figure 2.14 has a long right hand tail.

(b) The ordered data are

0 0 0 0 0 1 1 1 2 2 3

The median is the middle value, 1 car, and $\overline{x} = 10/11 = .909$ car which is smaller than the median.

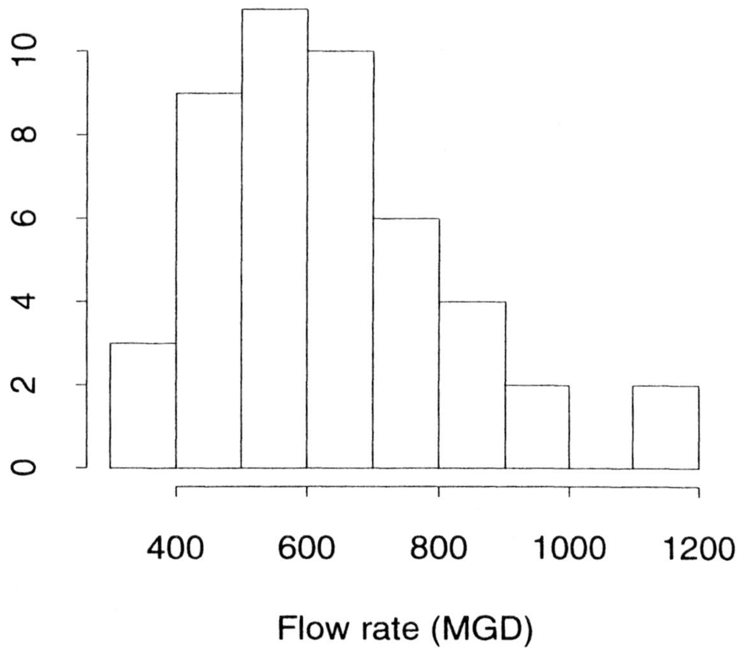

Figure 2.13: Histogram for Exercise 2.75(c).

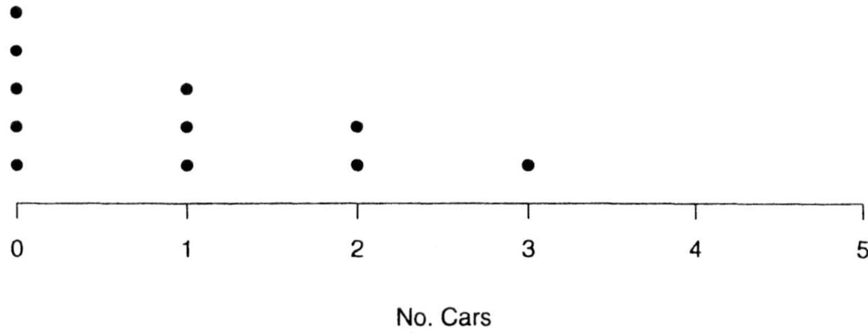

Figure 2.14: Dot diagram for Exercise 2.79(a).

Chapter 3

PROBABILITY

3.1 (a) A sketch of the 12 points of the sample space is as follows:

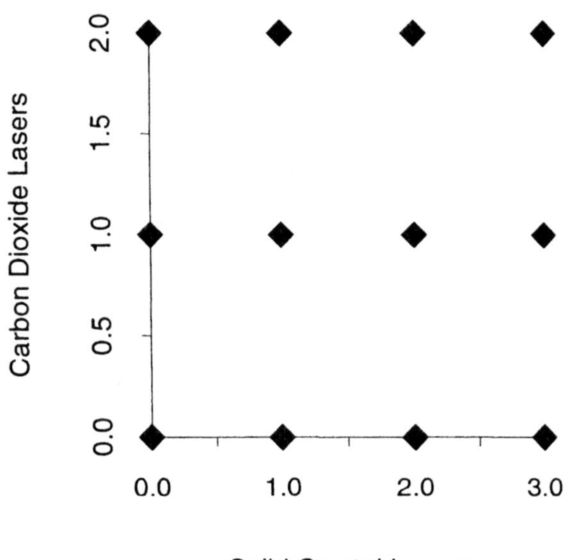

(b) R={(0,0),(1,1),(2,2)}. T={(0,0),(1,0),(2,0),(3,0)}. U={(0,1),(0,2),(1,2)}.

3.3 (a) $R \cup U$={(0,0),(1,1),(2,2),(0,1),(0,2),(1,2)}. $R \cup U$ is the event that the number of suitable carbon dioxide lasers is greater than or equal to the number of suitable solid crystal lasers.

(b) $R \cap T = \{(0,0)\}$. $R \cap T$ is the event that neither carbon dioxide lasers nor solid crystal lasers are suitable.

(c) $\overline{T} = \{(0,1),(1,1),(2,1),(3,1),(0,2),(1,2),(2,2),(3,2)\}$. $\overline{T}$ is the event that at least one carbon dioxide laser is suitable.

3.5 $A=\{3,4\}$, $B=\{2,3\}$, $C=\{4,5\}$.

(a) $A \cup B = \{2,3,4\}$. Work is easy, average or difficult on this model.

(b) $A \cap B = \{3\}$. Work is average on this model.

(c) $\overline{B} = \{1,4,5\}$. Thus $A \cup \overline{B} = \{1,3,4,5\}$. Work is not easy on this model.

(d) $\overline{C} = \{1,2,3\}$. Work is very easy, easy or average on this model.

3.7 (a) A sketch of the 6 points of the sample space is given in Figure 3.1.

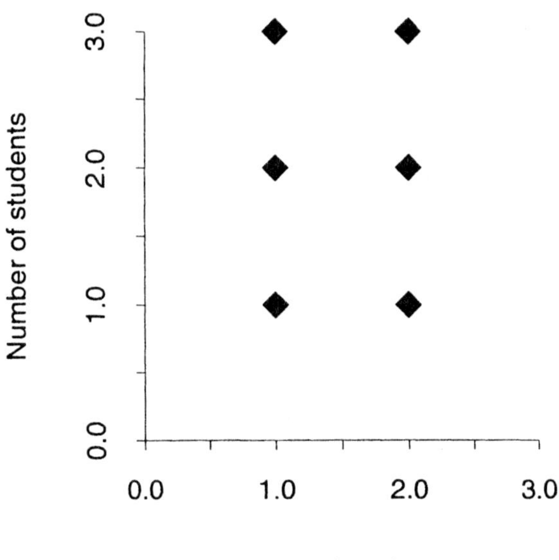

Figure 3.1: Sample space for Exercise 3.7.

(b) B is the event that 3 graduate students are present. C is the event that the same number of professors and graduate students are present. D is the event that the total number of graduate students and professors is 3.

(c) $C \cup D$={(1,1),(1,2),(2,1),(2,2)}. $C \cup D$ is the event that at most 2 graduate students are present.

(d) B and D are mutually exclusive.

3.9 Region 1 is the event that the ore contains both uranium and copper. Region 2 is the event that the ore contains copper but not uranium. Region 3 is the event that the ore contains uranium but not copper. Region 4 is the event that the ore contains neither uranium nor copper.

3.11 (a) Region 5 represents the event that the windings are improper, but the shaft size is not too large and the electrical connections are satisfactory.

(b) Regions 4 and 6 together represent the event that the electrical connections are unsatisfactory, but the windings are proper.

(c) Regions 7 and 8 together represent the event that the windings are proper and the electrical connections are satisfactory.

(d) Regions 1, 2, 3, and 5 together represent the event that the windings are improper.

3.13 The following Venn diagram will be used in parts (a), (b), (c) and (d).

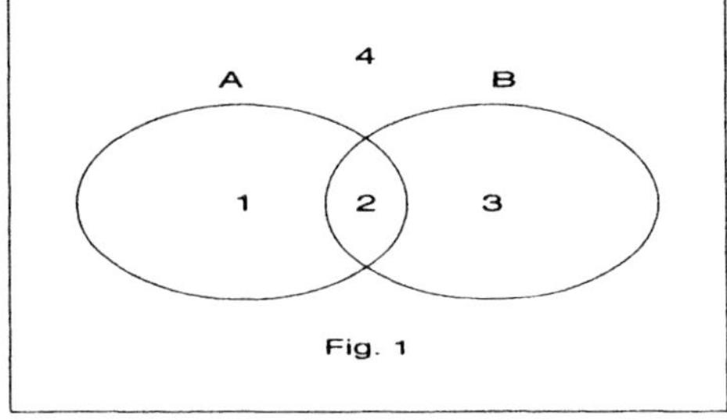

Fig. 1

(a) $A \cap B$ is region 2 in Fig. 1. $\overline{(A \cap B)}$ is the region composed of areas 1, 3, and 4. $\overline{A}$ is the region composed of areas 3 and 4. $\overline{B}$ is the region composed of areas 1 and 4. $\overline{A} \cup \overline{B}$ is the region composed of areas 1, 3, and 4. This corresponds to $\overline{(A \cap B)}$.

(b) $A \cap B$ is the region 2 in the figure. A is the region composed of areas 1 and 2. Since $A \cap B$ is entirely contained in A, $A \cup (A \cap B) = A$.

(c) $A \cap B$ is region 2. $A \cap \overline{B}$ is region 1. Thus, $(A \cap B) \cup (A \cap \overline{B})$ is the region composed of areas 1 and 2 which is A.

(d) From part (c), we have $(A \cap B) \cup (A \cap \overline{B}) = A$. Thus, we must show that $(A \cap B) \cup (A \cap \overline{B}) \cup (\overline{A} \cap B) = A \cup (\overline{A} \cap B) = A \cup B$. A is the region composed of areas 1 and 2 and $\overline{A} \cap B$ is region 3. Thus, $A \cup (\overline{A} \cap B)$ is the region composed of areas 1, 2, and 3.

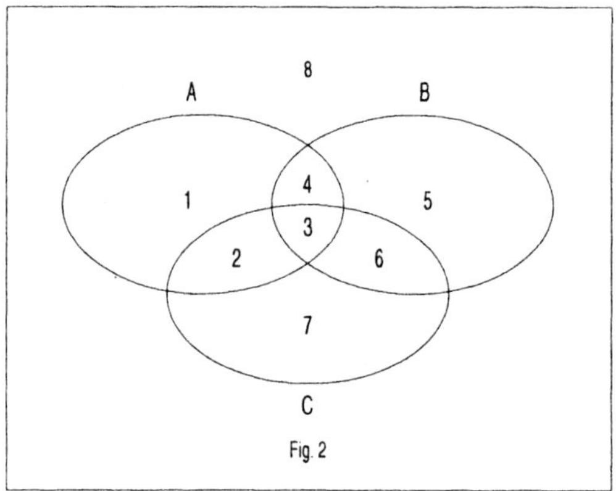

Fig. 2

(e) In Fig. 2, $A \cup B$ is the region composed of areas 1, 2, 3, 4, 5, and 6. $A \cup C$ is the region composed of areas 1, 2, 3, 4, 6, and 7, so $(A \cup B) \cap (A \cup C)$ is the region composed of areas 1, 2, 3, 4, and 6. $B \cap C$ is the region composed of areas 3, and 6, and A is the region composed of areas 1, 2, 3, and 4. Thus, $A \cup (B \cap C)$ is the region composed of areas 1, 2, 3, 4, and 6. Thus $A \cup (B \cap C) = (A \cup B) \cap (A \cup C)$.

3.15 The tree diagram is given in Figure 3.2, where S = Spain, U = Uruguay, P = Portugal and J = Japan.

3.17 There are $(6)(4)(3) = 72$ ways.

3.19 (a) ${}_8P_4 = 8!/4! = 1,680$.

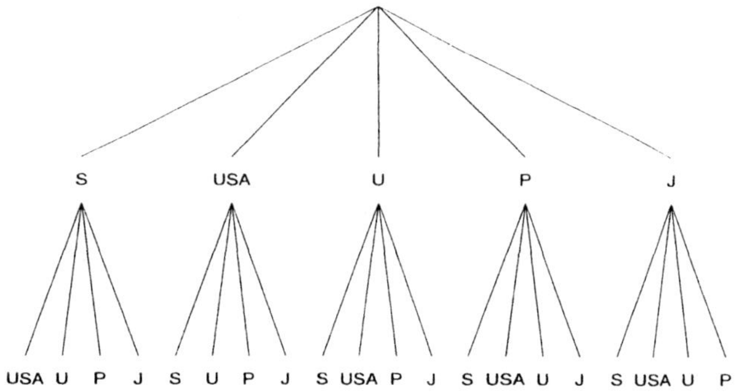

Figure 3.2: The tree diagram for Exercise 3.15.

(b) $8^4 = 4,096$.

3.21 $6! = 720$.

3.23 Since order does not matter, there are

$$_{15}C_2 = \binom{15}{2} = \frac{15!}{13!\,2!} = 105$$

ways.

3.25 There are $_{12}C_3 = 220$ ways to draw the three transistors.

There are $_{11}C_3 = 165$ ways to draw none are defective.

(a) The number of ways to get the one that is defective is $220 - 165 = 55$.

(b) There are 165 ways not to get the one that is defective.

3.27 There are $_8C_2$ ways to choose the refrigerators, $_6C_2$ ways to choose the washer-dryers, and $_5C_2$ ways to choose the microwaves. Thus, there are

$$_8C_2 \cdot {}_6C_2 \cdot {}_5C_2 = 28 \cdot 15 \cdot 10 = 4,200$$

ways to choose the appliances for the sale.

3.29 The outcome space is given in Figure 3.3.

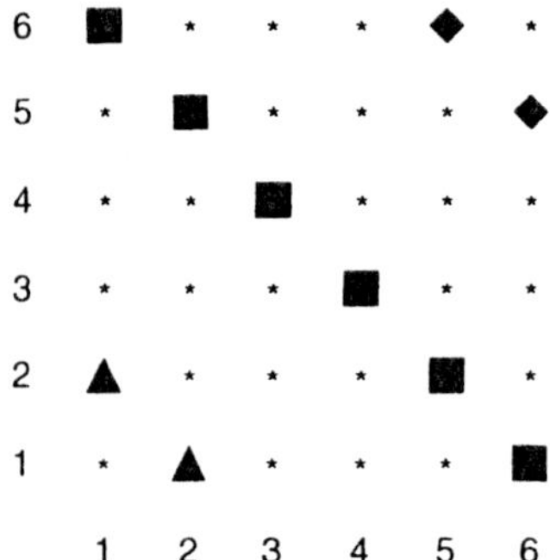

Figure 3.3: The outcome space for Exercise 3.29.

(a) The 6 outcomes summing to 7 are marked by squares. Thus, the probability is $6/36 = 1/6$.

(b) There are 2 outcomes summing to 11, which are marked by diamonds. Thus, the probability is $2/36 = 1/18$.

(c) These events are mutually exclusive. Thus, the probability is $6/36 + 2/36 = 2/9$.

(d) The 2 outcomes are marked by triangles. Thus, the probability is $2/36 = 1/18$.

(e) There are two such outcomes, (1,1) and (6,6). Thus, the probability is $2/36 = 1/18$.

(f) There are four such outcomes, (1,1), (1,2), (2,1) and (6,6). Thus, the probability is $4/36 = 1/9$.

3.31 There are 18 + 12 = 30 cars. Thus, there are ${}_{30}C_4$ ways to choose the cars for inspection. There are ${}_{18}C_2$ ways to get the compacts and ${}_{12}C_2$ ways to get the intermediates. Thus, the probability is:

$$\frac{\binom{18}{2}\binom{12}{2}}{\binom{30}{4}} = \frac{10,098}{27,405} = .368.$$

3.33 The number of students enrolled in the statistics course or the operations research course is $92 + 63 - 40 = 115$. Thus, $160 - 115 = 45$ are not enrolled in either course.

3.35 (a) Yes. $P(A) + P(B) + P(C) + P(D) = 1$.

(b) No. $P(A) + P(B) + P(C) + P(D) = 1.02 > 1$.

(c) No. $P(C) = -.06 < 0$.

(d) No. $P(A) + P(B) + P(C) + P(D) = 15/16 < 1$.

(e) Yes. $P(A) + P(B) + P(C) + P(D) = 1$.

3.37 (a) There is 1 point where $i + j = 2$. There are 2 points where $i + j = 3$. There are 2 points where $i + j = 4$. There is 1 point where $i + j = 5$. Thus,

$$\begin{aligned} &\frac{15/28}{2} + 2 \cdot \frac{15/28}{3} + 2 \cdot \frac{15/28}{4} + \frac{15/28}{5} \\ &= \frac{15}{28} \cdot (1/2 + 2/3 + 1/2 + 1/5) = \frac{15}{28} \cdot \frac{28}{15} = 1. \end{aligned}$$

Since each probability is between 0 and 1, the assignment is permissible.

(b) $P(B) = 15/28 \cdot (1/4 + 1/5) = 5/28 \cdot 9/20 = 135/560 = 27/112$.
$P(C) = 15/28 \cdot (1/2 + 1/4) = 15/28 \cdot 3/4 = 45/112$.
$P(D) = 15/28 \cdot (1/3 + 1/3) = 15/28 \cdot 2/3 = 5/14$.

(c) The probability that 1 graduate student will be supervising the lab is:

$$\frac{15/28}{2} + \frac{15/28}{3} = \frac{25}{56} = .446.$$

The probability that 2 graduate students will be supervising the lab is:

$$\frac{15/28}{3} + \frac{15/28}{4} = \frac{35}{112} = \frac{5}{16} = .3125.$$

The probability that 3 graduate students will be supervising the lab is:

$$\frac{15/28}{4} + \frac{15/28}{5} = \frac{27}{112} = .241.$$

3.39 (a) $(A \cap B) \cup (A \cap \overline{B}) = A$, and $A \cap B$ and $A \cap \overline{B}$ are disjoint. Thus

$$P((A \cap B) \cup (A \cap \overline{B})) = P(A \cap B) + P(A \cap \overline{B}) = P(A).$$

Since $P(A \cap \overline{B}) \geq 0$, we have proved that $P(A \cup B) \leq P(A)$.

(b) Combining (d) and (c) of Exercise 3.13 gives us

$$A \cup B = (A \cap B) \cup (A \cap \overline{B}) \cup (\overline{A} \cap B) = A \cup (\overline{A} \cap B).$$

But A and $\overline{A} \cap B$ are disjoint. Thus,

$$P(A \cup B) = P(A) + P(\overline{A} \cap B).$$

Since $P(\overline{A} \cap B) \geq 0$, we have proved that $P(A \cup B) \geq P(A)$.

3.41 (a) $P(\overline{A}) = 1 - P(A) = 1 - .29 = .71$.

(b) $P(A \cup B) = P(A) + P(B) = .29 + .43 = .72$, since A and B are mutually exclusive.

(c) $P(A \cap \overline{B}) = P(A) = .29$, since A and B are mutually exclusive.

(d) $P(\overline{A} \cap \overline{B}) = P(\overline{(A \cup B)}) = 1 - P(A \cap B) = 1 - .29 - .43 = .28$.

3.43 (a) This probability is given by $.22 + .21 = .43$.

(b) $.17 + .29 + .21 = .67$.

(c) $.03 + .08 = .11$.

(d) $.22 + .29 + .08 = .59$.

3.45 (a) 15/32 (b) 13/32 (c) 5/32 (d) 23/32 (e) 8/32 (f) 9/32.

3.47 (a) "At least one award" is the same as "design or efficiency award". Thus, the probability is $.16 + .24 - .11 = .29$.

(b) This the probability of "at least one award" minus the probability of both awards or $.29 - .11 = .18$.

3.49

$$P(A \cup B \cup C) = 1 - .11 = .89, \qquad P(A) = .24 + .06 + .04 + .16 = .5,$$
$$P(B) = .19 + .06 + .04 + .11 = .4, \qquad P(C) = .09 + .16 + .04 + .11 = .4,$$
$$P(A \cap B) = .06 + .04 = .1, \qquad P(A \cap C) = .16 + .04 = .2,$$
$$P(B \cap C) = .04 + .11 = .15, \qquad P(A \cap B \cap C) = .04.$$

Thus, the following equation must equal to .89:

$$.5 + .4 + .4 - .1 - .2 - .15 + .04 = .89.$$

This proves the formula.

3.51 (a) The odds for are $(4/7)/(3/7) = 4$ to 3.

(b) The odds against are $.95/.05 = 19$ to 1 against.

(c) The odds for are $.80/.20 = 4$ to 1.

3.53 (a) $p = 3/(3+2) = 3/5$.

(b) $30/(30+10) = 3/4 \leq p < 40/(10+40) = 4/5$.

tem[3.55] $P(I \cap D) = 10/500$, $P(D) = 15/500$, $P(I \cap \overline{D}) = 20/500$, $P(\overline{D}) = 485/500$.

$$P(I|D) = \frac{P(I \cap D)}{P(D)} = \frac{10}{15} = \frac{2}{3},$$

$$P(I|\overline{D}) = \frac{P(I \cap \overline{D})}{P(\overline{D})} = \frac{20}{485} = \frac{4}{97}.$$

3.57 (a) The sample space is C. Thus the probability is given by:

$$\frac{N(A \cap C)}{N(C)} = \frac{8+15}{8+54+9+14} = \frac{62}{85} = .73.$$

(b) This is given by:

$$\frac{N(A \cap B)}{N(A)} = \frac{20+54}{20+54+8+2} = \frac{74}{84} = .881.$$

(c) This is given by:

$$\frac{N(\overline{C} \cap \overline{B})}{N(\overline{B})} = \frac{N(\overline{(C \cup B)})}{N(\overline{B})}. = \frac{150-121}{105-99} = \frac{29}{51} = .569.$$

3.59 (a) $P(P_1) = (.07 + .13 + .06) + (.07 + .14 + .07) + (.02 + .03 + .01) = .6,$

$P(E_1) = .07 + .13 + .06 + .05 + .07 + .02 = .4,$

$P(E_1 \cap P_1) = .07 + .13 + .06 =.26.$

Hence, $P(E_1|P_1) = .26/.6 = .433$. And $P(E_1|P_1) > P(E_1)$.

(b) $P(P_2) = (.05 + .07 + .02) + (.08 + .11 + .03) + (.01 + .02 + .01) = .4,$

$P(C_2) = .13 + .07 + .14 + .11 + .03 + .02 = .5,$

$P(C_2 \cap P_2) = .07 + .11 + .02 =.2.$

Hence, $P(C_2|P_2) = .2/.4 = .5$. And $P(C_2|P_2) = P(C_2)$.

(c) $P(E_1) = .4,$

$P(P_1 \cap C_1) = .07 + .07 + .02 =.16,$

$P(E_1 \cap P_1 \cap C_1) = .07.$

Hence, $P(E_1|P_1 \cap C_1) = .07/.16 = .4375$. And $P(E_1|P_1 \cap C_1) > P(E_1)$.

3.61 $P(A|B) = P(A \cap B)/P(B)$ by definition. Thus, $P(A|B) = P(A)$ implies that $P(A \cap B)/P(B) = P(A)$, which implies $P(A \cap B)/P(A) = P(B)$, since both $P(A)$ and $P(B)$ are not zero. Thus $P(B|A) = P(B)$.

3.63 (a) $P(A|B) = P(A \cap B)/P(B) = .15/.30 = .5 = P(A)$.

(b) $P(A|\overline{B}) = P(A \cap \overline{B})/P(\overline{B}) = (P(A) - P(A \cap B))/(1 - P(B))$
$= (.50 - .15)/(1 - .30) = .5 = P(A)$.

(c) $P(B|A) = P(B \cap A)/P(A) = .15/.50 = .3 = P(B)$.

(d) $P(B|\overline{A}) = P(B \cap \overline{A})/P(\overline{A}) = (P(B) - P(B \cap A))/(1 - P(A))$
$= (.30 - .15)/(1 - .50) = .3 = P(B)$.

3.65 (a) The probability of drawing a Seattle-bound part on the first draw is 45/60. The probability of drawing a Seattle-bound part on the second draw given that a Seattle-bound part was drawn on the first draw is 44/59. Thus, the probability that both parts should have gone to Seattle is:

$$\frac{45}{60} \cdot \frac{44}{59} = .559.$$

(b) Using an approach similar to (a), the probability that both parts should have gone to Vancouver is:

$$\frac{15}{60} \cdot \frac{14}{59} = .059.$$

(c) The probability that one should have gone to Seattle and one to Vancouver is 1 minus the sum of the probability in parts (a) and (b) or .381.

3.67 A and C are independent if and only if

$$P(A)P(C) = P(A \cap C).$$

Since (.8)(.35)=.28, they are independent.

3.69 (a) Each head has probability 1/2, and each toss is independent. Thus, the probability of 8 heads is $(1/2)^8 = 1/256$.

(b) $P(\text{three 3's and then a 4 or 5}) = (1/6)^3(1/3) = 1/648$.

(c) $P(\text{five questions answered correctly}) = (1/3)^5 = 1/243$.

3.71 $P(\text{new worker meets quota}) = (.80)(.86) + (.20)(.35) = .758$.

3.73 $P(\text{car had bad tires}) = (.20)(.10)+(.20)(.12)+(.60)(.04)=.068$.

3.75 (a) $P(A) = (.4)(.3) + (.6)(.8) = .60$.

(b) $P(B|A) = P(B \cap A)/P(A) = (.4)(.3)/(.60) =.20$.

(c) $P(B|\overline{A}) = P(B \cap \overline{A})/P(\overline{A}) = (.4)(.7)/(.4) =.70$.

3.77 (a)

$$\begin{aligned} &P(\text{Tom} \mid \text{not worked}) \\ &\quad = \frac{(.6)(1/10)}{(.2)(1/20) + (.6)(1/10) + (.15)(1/10) + (.05)(1/20)} \\ &\quad = \frac{.06}{.0875} = .686. \end{aligned}$$

(b)

$$P(\text{George} \mid \text{not worked}) = \frac{(.15)(1/10)}{.0875} = .171.$$

(c)

$$P(\text{Peter} \mid \text{not worked}) = \frac{(.05)(1/20)}{.0875} = .0286.$$

3.79 Let A be the event that the test indicates corrosion inside of the pipe and C be the event that corrosion is present. We are given $P(A|C) = .7$, $P(A|\overline{C}) = .2$, and $P(C) = .1$.

(a) By Bayes' theorem

$$P(C|A) = \frac{P(A|C)P(C)}{P(A|C)P(C) + PA|\overline{C})P(\overline{C})}$$

$$= \frac{.7 \times .1}{.7 \times .1 + .2 \times .9} = \frac{.07}{.07 + .18} = .28$$

(b)

$$P(C|\overline{A}) = \frac{P(\overline{A}|C)P(C)}{P(\overline{A}|C)P(C) + P(\overline{A}|\overline{C})P(\overline{C})}$$

$$= \frac{(1 - .7) \times .1}{(1 - .7) \times .1 + (1 - .2) \times .9} = \frac{.03}{.03 + .72} = .04$$

3.81 The expected gain is (800)(1/4,000) + (0)(3,999/4,000) = \$.20.

3.83 The expected gain is (10)(1/2) + (−10)(1/2) = 0 dollars.

3.85 The expected length of a World Series is

$$4(1/8) + 5(1/4) + 6(5/16) + 7(5/16) = 93/16.$$

3.87 If 6 items are stocked, the cost is (6)(35)=\$210. The expected revenue is:

$$(3.50)(.05) + (4.50)(.12) + (5.50)(.20) + (6.50)(.24 + .17 + .14 + .08) = \$270.5.$$

Thus, the expected profit is \$270.5−\$210=\$60.50.

If 7 items are stocked, the cost is (7)(35)=\$245. The expected revenue is:

$$(3.50)(.05) + (4.50)(.12) + (5.50)(.20) + (6.50)(.24) + (7.50)(.17 + .14 + .08) = \$290.$$

Thus, the expected profit is \$290−\$245=\$45.

3.89 (a) He feels that $p(-1.20) + (1 - p)(1.20) < .80$, where p is the probability that a person will ask for double his money back. Thus, $p > 1/6$.

(b) Changing the direction of the inequalities in part (a) gives $p < 1/6$.

(c) Changing the inequalities in part (a) to equalities gives p=1/6.

3.91 Brown's expected gain is given by $pn + (1-p)(-m)$ which must be 0 since the game is fair. Thus, $p = m/(n+m)$. If Brown begins with \$12 and Jones with \$4, the probability that Brown wins is 12/(12 + 4)=3/4.

3.93 Now, the expected gain if they continue is

$$(\$1,000,000)(.20) - (\$600,000)(.80) = -\$280,000.$$

If they do not continue, the expected gain is

$$(-\$400,000)(.20) + (\$100,000)(.80) = \$0.$$

They should not continue the operation.

3.95 (a) Using the long run relative frequency approximation to probability, we estimate the probability

$$P[\,\text{Checked out}\,] = \frac{27}{300} = 0.09$$

(b) Using the data from last year, the long run relative frequency approximation to probability gives the estimate

$$P[\,\text{Get internship}\,] = \frac{28}{380} = 0.074$$

One factor might be the quality of permanent jobs that interns received last year or even how enthusiastic they were about the internship. Both would likely increase the number of applicants. Bad experiences may decrease the number of applicants.

3.97 (a) $\overline{A}$={(0,0),(0,1),(0,2),(0,3),(1,0),(1,1),(1,2),(2,0), (2,1),(3,0)}.

$\overline{A}$ is the event that the salesman will not visit all four of his customers.

(b) $A \cup B$={(4,0),(3,1),(2,2),(1,3),(0,4),(1,0),(2,0), (2,1),(3,0)}.

$A \cup B$ is the event that the salesman will visit all four customers or more on the first day than on the second day.

(c) $A \cap C$={(1,3),(0,4)}.

$A \cap C$ is the event that he will visit all four customers but at most one on the first day.

(d) $\overline{A} \cap B$={(1,0),(2,0),(2,1),(3,0)}.

$\overline{A} \cap B$ is the event that he will visit at most three of the customers and more on the first day than on the second day.

3.99 The tree diagram is given in Figure 3.4.

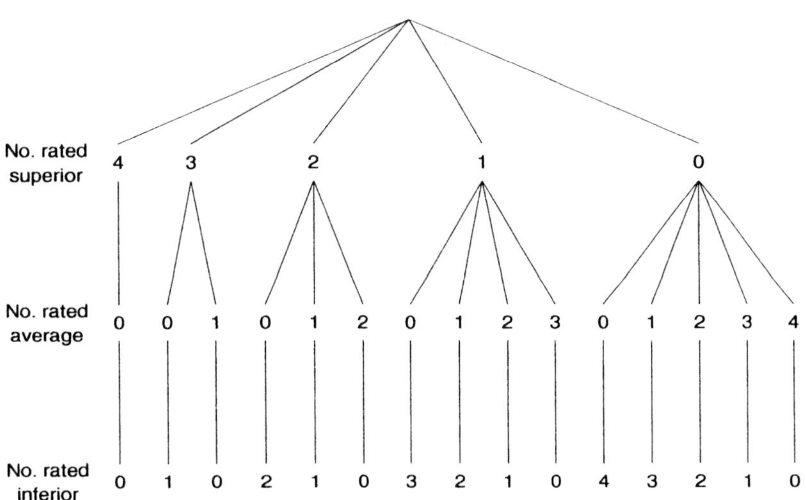

Figure 3.4: Tree diagram for Exercise 3.99.

3.101 There are ${}_7C_2 = 21$ ways to assign the chemical engineers.

3.103 The Venn diagram is given in Figure 3.5.

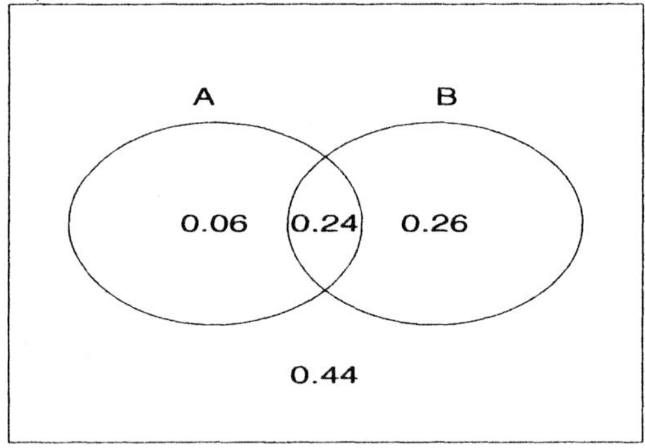

Figure 3.5: Venn diagram for Exercise 3.103.

(a) $P(A \cup B) = .06 + .24 + .26 = .56.$

(b) $P(\overline{A} \cap B) = .26.$

(c) $P(A \cap B) = .06.$

(d) $P(\overline{A} \cup \overline{B}) = P(\overline{(A \cap B)}) = 1 - P(A \cap B) = 1 - .24 = .76.$

(e) A and B are not independent, since

$$P(A)P(B) = (.3)(.5) = .15 \neq P(A \cap B).$$

3.105 The total number of employees that got a raise or an increase in pension benefits is $312 + 248 - 173 = 387$. If 43 got neither, then there must be $387 + 43 = 430$ employees, not 400 employees.

3.107 (a) $P(A|B) = P(A \cap B)/P(B) = .2/.5 = .4 = P(A).$

(b) Since $P(A \cap \overline{B}) = P(A) - P(A \cap B) = .4 - .2 = .2$, hence

$$P(A|\overline{B}) = P(A \cap \overline{B})/P(\overline{B}) = .2/.5 = .4 = P(A).$$

(c) $P(B|A) = P(B \cap A)/P(A) = .2/.4 = .5 = P(B)$.

(d) Since $P(B \cap \overline{A}) = P(B) - P(B \cap A) = .5 - .2 = .3$, hence

$$P(B|\overline{A}) = P(B \cap \overline{A})/P(\overline{A}) = .3/(1 - .4) = .5 = P(B).$$

3.109 Let us denote the chemicals Arsenic, Barium, and Mercury by the letters A, B, and M respectively, and indicate the concentrations by the subscripts 'H' for high and 'L' for low. For instance, a high concentration of Mercury will be denoted by M_H.

(a) Of the 58 landfills, the number with M_H is $1 + 4 + 5 + 10 = 20$.

Therefore, $P(M_H) = \frac{20}{58} = .344$

(b) The number of $M_L A_L B_H$ landfills is 8 so $P(M_L A_L B_H) = \frac{8}{58} = .138$

(c) There are three possibilities for landfills with two H's one L. The number of $A_H B_L M_H$ landfills is 5, the number of $A_L B_H M_H$ is 4, and the number of $A_H B_H M_L$ is 3, so the total is 12.

Therefore, $P(\text{two } H\text{'s and one } L\text{'s}) = \frac{12}{58} = .207$

(d) There are three possibilities for landfills with one H and two L's. The number of $A_H B_L M_L$ landfills is 9, the number of $A_L B_H M_L$ is 8, and the number of $A_L B_L M_H$ is 10, so the total is 27.

Therefore, $P(\text{one } H \text{ and two } L\text{'s}) = \frac{27}{58} = .466$

3.111 Let events S.E. = static electricity, E = explosion, M = malfunction, O.F. = open flame, and P.E. = purposeful action. We need to find probabilities $P(\text{S.E.}|\text{E})$, $P(\text{M}|\text{E})$, $P(\text{O.F.}|\text{E})$, $P(\text{P.A.}|\text{E})$. Since

$$P(E) = (.30)(.25) + (.40)(.20) + (.15)(.40) + (.15)(.75) = .3275,$$

we have

$$P(S.E.|E) = (.30)(.25)/.3275 = .229, \qquad P(M|E) = (.40)(.20)/.3275 = .244,$$
$$P(O.F.|E) = (.15)(.40)/.3275 = .183, \qquad P(P.A.|E) = (.15)(.75)/.3275 = .344.$$

Thus, purposeful action is most likely.

3.113 The expected profit on the first job is

$$(.75)(240,000) + (.25)(-60,000) = \$165,000.$$

The expected profit on the second job is

$$(.5)(360,000) + (.5)(-90,000) = \$135,000.$$

(a) He should choose the first job.

(b) Only if he chooses the second job does he have any chance of not going broke.

3.115 We let A be the event route A is selected, B be the event route B is selected and C the event Amy arrives home at or before 6 p.m. We are given $P(A) = .4$ so $P(B) = 1 - .4 = .6$.

(a) By the law of total probability

$$P(C) = P(C|A)P(A) + P(C|B)P(B) = .8 \times .4 + \ .7 \times .6 = .74$$

(b) Using Bayes' rule

$$P(B|\overline{C}) = \frac{P(\overline{C}|B)P(B)}{P(\overline{C}|A)P(A) + P(\overline{C}|B)P(B)} = \frac{.3 \times .6}{.2 \times .4 + .3 \times .6} = .692$$

Chapter 4

PROBABILITY DISTRIBUTIONS

4.1 Let N be the number of suitable lasers.

$$P(N=0) = 1/2 \qquad P(N=3) = 3/12 = 1/4$$
$$P(N=1) = 2/12 = 1/6 \qquad P(N=4) = 2/12 = 1/6$$
$$P(N=2) = 3/12 = 1/4 \qquad P(N=5) = 1/12.$$

Thus, the distribution can be tabulated as :

N	0	1	2	3	4	5
Prob	1/12	1/6	1/4	1/4	1/6	1/12

4.3 (a) No. $\sum_{i=1}^{4} f(i) \;=\; 1.04 > 1.$

(b) Yes. $0 \le f(i) \le 1$, and $\sum_{i=1}^{4} f(i) \;=\; 1.$

(c) No. $f(3) < 0.$

4.5 Using the identity

$$(x-1)\sum_{i=0}^{n} x^i \;=\; x^{n+1} - 1$$

or

$$\sum_{i=0}^{n} x^i = \frac{x^{n+1}-1}{x-1},$$

we have

$$\sum_{x=0}^{4} \frac{k}{2^x} = k\frac{(\frac{1}{2})^{4+1}-1}{\frac{1}{2}-1} = \frac{31k}{16}.$$

This must equal 1, so $k = 16/31$.

4.7

$$\begin{aligned} b(x;n,p) &= \binom{n}{x} p^x(1-p)^{n-x} = \frac{n!}{x!\,(n-x)!}\; p^x(1-p)^{n-x} \\ &= \frac{n!}{(n-x)!\,x!}\;(1-p)^{n-x}\,p^x = \binom{n}{n-x}(1-p)^{n-x}\,p^x \\ &= b(n-x;n,1-p) \end{aligned}$$

4.9 (a) Assumptions appear to hold. Success is a home with a TV tuned to mayor's speech. The probability of success is the proportion of homes around city having a TV tuned to the mayor's speech.

(b) The binomial assumptions do not hold because the probability of a serious violation for the second choice depends on which plant is selected first.

4.11 (a) Success is person has a cold. Colds are typically passed around in families so trials would not be independent. Therefore, the binomial distribution does not apply.

(b) Success means projector does not work properly. The binomial assumptions do not hold because the probability of a success for the second choice depends on which projector is selected first.

4.13 From Table 1:

(a) $B(7;19,.45) = .3169$.

(b) $b(7;19,.45) = (.3169) - (.1727) = .1442.$

(c) $B(8;10,.95) = .0861.$

(d) $b(8;10,.95) = (.0861) - (.0115) = .0746.$

(e) $\sum_{k=4}^{10} b(k;10,.35) = 1 - B(3;10,.35) = 1 - (.5138) = .4862.$

$B(4;9,.3) - B(1;9,.3) = (.9012) - (.1960) = .7052.$

4.15

$$b(2;4,.75) = \binom{4}{2}(.75)^2(.25)^{4-2} = .2109.$$

4.17 (a) $1 - B(11;\ 15,\ .7) = 1 - .7031 = .2969.$

(b) $B(6;\ 15,\ .7) = .0152.$

(c) $b(10;\ 15,\ .7) = B(10;\ 15,\ .7) - B(9;\ 15,\ .7) = (.4845) - (.2784) = .2061.$

4.19 (a) $P(18 \text{ are ripe}) = (.9)^{18} = .1501.$

(b) $1 - B(15;\ 18,\ .9) = 1 - .2662 = .7338.$

(c) $B(14;\ 18,\ .9) = .0982.$

4.21 (a) $B(2;\ 16,\ .05\ ) = .9571$

(b) $B(2;\ 16,\ .10\ ) = .7892$

(c) $B(2;\ 16,\ .15\ ) = .5614$

(d) $B(2;\ 16,\ .20\ ) = .3518$

4.23 For this problem, we need to use the hypergeometric distribution. The probability is given by:

$$h(2;6,8,18) = \frac{\binom{8}{2}\binom{10}{4}}{\binom{18}{6}} = \frac{(8!)\ (10!)\ (12!)}{2\ (6!)\ (4!)\ (18!)} = .3167$$

4.25 Using the hypergeometric distribution,

(a)

$$h(0;3,6,24)=\frac{\binom{6}{0}\binom{18}{3}}{\binom{24}{3}}=\frac{816}{2024}=.4032$$

(b)

$$h(1;3,6,24)=\frac{\binom{6}{1}\binom{18}{2}}{\binom{24}{3}}=\frac{6\cdot 153}{2024}=.4536$$

(c) The probability that at least 2 have defects is 1 minus the sum of the probabilities of none, and 1 having defects.

$$1-(.4032+.4536)=.1432$$

4.27 (a)

$$P(\text{none in west})=h(0;3,7,16)=\frac{\binom{7}{0}\binom{9}{3}}{\binom{16}{3}}=\frac{1\cdot 84}{560}=.15$$

(b)

$$P(\text{all in west}) = h(3;3,7,16) = \frac{\binom{7}{3}\binom{9}{0}}{\binom{16}{3}} = \frac{35 \cdot 1}{560} = .0625$$

4.29 (a)

$$P(5 \text{ are union members}) = \frac{\binom{240}{5}\binom{60}{3}}{\binom{300}{8}} = .1470$$

(b)

$$P(5 \text{ are union members}) = \binom{8}{5}\left(\frac{240}{300}\right)^4\left(\frac{60}{300}\right)^3 = .1468$$

4.31 The cumulative binomial probabilities are

```
CDF;
BINOMIAL n = 27 p = .47 .

BINOMIAL WITH N =   27   P = 0.470000
   K  P( X LESS OR = K)
   2            0.0000
   3            0.0001
   4            0.0005
   5            0.0021
   6            0.0072
   7            0.0210
   8            0.0515
```

9	0.1086
10	0.1998
11	0.3247
12	0.4724
13	0.6236
14	0.7576
15	0.8607
16	0.9292
17	0.9685
18	0.9879
19	0.9960
20	0.9989
21	0.9997
22	1.0000

4.33 Using the computing formula:

$$\begin{aligned} \sigma^2 &= \mu_2' - \mu^2 \quad \text{with} \quad \mu = 1 \\ \mu_2' &= 0^2(.4) + 1^2(.3) + 2^2(.2) + 3^2(.1) = 2 \end{aligned}$$

Thus,

$$\sigma^2 = 2 - 1^2 = 1.$$

4.35 Using the computing formula:

$$\sigma^2 = \mu_2' - \mu^2, \qquad \mu = 1.8$$

$$\mu_2' = 0^2(.17) + 1^2(.29) + 2^2(.27) + 3^2(.16) + 4^2(.07) + 5^2(.03) + 6^2(.01) = 5.04$$

Thus,

$$\sigma^2 = 5.04 - (1.8)^2 = 1.8.$$

The standard deviation is $\sigma = \sqrt{1.8} = 1.34$.

4.37 (a) The mean for the binomial distribution with $n = 4$ and $p = 7$ can be calculated from the following table:

i	0	1	2	3	4
$b(i; 4, .7)$	.0081	.0756	.2646	.4116	2401

Thus,

$$\begin{aligned}\mu &= 0(.0081) + 1(.0756) + 2(.2646) + 3(.4116) + 4(.2401) = 2.8 \\ \mu_2' &= 0^2(.0081) + 1^2(.0756) + 2^2(.2646) + 3^2(.4116) + 4^2(.2401) = 8.68 \\ \sigma^2 &= 8.68 - (2.8)^2 = .84.\end{aligned}$$

(b) $\mu = np = 4(.7) = 2.8.$

$\sigma^2 = np(1-p) = 4(.7\ )(.3) = .84.$

4.39 (a) The variance is given by:

$$\sigma^2 = (0-2.5)^2\frac{1}{32} + (1-2.5)^2\frac{5}{32} + (2-2.5)^2\frac{10}{32} + (3-2.5)^2\frac{10}{32}$$

$$+(4-2.5)^2\frac{5}{32} + (5-2.5)^2\frac{1}{32} = \frac{40}{32} = 1.25$$

(b) To use the computing formula we need:

$$\mu_2' = 0^2 \cdot \frac{1}{32} + 1^2 \cdot \frac{5}{32} + 2^2 \cdot \frac{10}{32} + 3^2 \cdot \frac{10}{32} + 4^2 \cdot \frac{5}{32} + 5^2 \cdot \frac{1}{32} = \frac{240}{32} = 7.5$$

Thus

$$\sigma^2 = 7.5 - (2.5)^2 = 1.25.$$

(c) The special formula for the binomial variance is:

$$\sigma^2 = np(1-p) = 5(.5)(.5) = 1.25$$

4.41 Since all of these random variables have binomial distributions,

(a)

$$\mu = np = 676(.5) = 338$$

$$\sigma^2 = np(1-p) = 676(.5)(.5) = 169$$

$$\sigma = 13$$

(b)

$$\mu = np = 720 \cdot \frac{1}{6} = 120$$

$$\sigma^2 = np(1-p) = 720 \cdot \frac{1}{6}\frac{5}{6} = 100$$

$$\sigma = 10$$

(c)

$$\mu = np = 600(.04) = 24$$

$$\sigma^2 = np(1-p) = 600(.04)(.96) = 23.04$$

$$\sigma = 4.8$$

(d)

$$\mu = np = 800(.65) = 520$$

$$\sigma^2 = np(1-p) = 800(.65)(.35) = 182$$

$$\sigma = 13.49$$

4.43 The mean of the hypergeometric distribution is

$$\mu = \sum_{x=0}^{n} x \frac{\binom{a}{x}\binom{N-a}{n-x}}{\binom{N}{n}}$$

$$= \sum_{x=1}^{n} \frac{x\binom{a}{x}\binom{N-a}{n-x}}{\binom{N}{n}}$$

$$= \frac{a}{\binom{N}{n}} \sum_{x=1}^{n} \binom{a-1}{x-1}\binom{N-a}{n-x}$$

Let $u = x - 1$. Then,

$$\mu = \frac{a}{\binom{N}{n}} \sum_{u=0}^{n-1} \binom{a-1}{u}\binom{N-a}{n-1-u}$$

Using the identity given in the problem, we have

$$\mu = \frac{a\binom{N-1}{n-1}}{\binom{N}{n}} = \frac{an}{N}$$

The result holds for all n such that $0 \le n \le N$, because

$$\binom{a}{x}\binom{N-a}{n-x} = 0 \quad \text{if} \quad x > a \quad \text{or} \quad (n-x) > (N-a)$$

4.45 Since $\sigma = 0.002$ we have $0.006 = 3(0.0002)$, and $k = 3$.

$$P(\mid X - \mu \mid \le 0.006)1 \; - \; P(\mid X - \mu \mid > 0.006) \ge 1 \; - \; \frac{1}{3^2} = \frac{8}{9}$$

4.47

$$\mu = 1,000,000 \cdot \frac{1}{2} = 500,000.$$

$$\sigma^2 = 1,000,000 \cdot \frac{1}{2} \cdot \frac{1}{2} = 250,000 \quad , \quad \sigma = 500.$$

If the proportion is between .495 and .505, the number of heads must be between 495,000 and 505,000. These bounds are both within 10 standard deviations of the mean. We can apply Chebyshev's theorem with $k = 10$. Thus, the probability is greater than $1 - 1/100 = .99$.

4.49 (a) By definition of variance,

$$\begin{aligned}
\sigma^2 &= \sum_{\text{all } x} (x-\mu)^2 f(x) \\
&= \sum_{\text{all } x} (x^2 - 2x\mu + \mu^2) f(x) \\
&= \sum_{\text{all } x} x^2 f(x) - \sum_{\text{all } x} 2x\mu f(x) + \sum_{\text{all } x} \mu^2 f(x) \\
&= \mu_2' - 2\mu \sum_{\text{all } x} x f(x) + \mu^2 \sum_{\text{all } x} f(x) \\
&= \mu_2' - 2\mu \cdot \mu + \mu^2 \cdot 1 \\
&= \mu_2' - \mu^2
\end{aligned}$$

(b) Similarly,

$$\begin{aligned}
\mu_3 &= \sum_{\text{all } x} (x-\mu)^3 f(x) \\
&= \sum_{\text{all } x} (x^3 - 3x^2\mu + 3x\mu^2 - \mu^3) f(x) \\
&= \sum_{\text{all } x} x^3 f(x) - \sum_{\text{all } x} 3x^2\mu f(x) + \sum_{\text{all } x} 3x\mu^2 f(x) - \sum_{\text{all } x} \mu^3 f(x) \\
&= \mu_3' - 3\mu \sum_{\text{all } x} x^2 f(x) + 3\mu^2 \sum_{\text{all } x} x f(x) - \mu^3 \sum_{\text{all } x} f(x) \\
&= \mu_3' - 3\mu \cdot \mu_2' + 3\mu^2 \cdot \mu - \mu^3 \cdot 1 \\
&= \mu_3' - 3\mu\mu_2' + 2\mu^3
\end{aligned}$$

4.51 For $\lambda = 3$, $f(0; \lambda) = e^{-3} = .0498$. Thus, using

$$f(x+1; \lambda) = \frac{\lambda}{x+1} f(x; \lambda)$$

$$f(1; \lambda) = \frac{\lambda}{0+1} f(0; \lambda) = 3e^{-3} = .1494.$$

$$f(2; \lambda) = \frac{3}{2} \cdot 3e^{-3} = .2240.$$

and so forth. The values are given in the following table:

x	0	1	2	3	4
$f(x;3)$	.0498	.1494	.2240	.2240	.1680

x	5	6	7	8	9
$f(x;3)$	.1008	.0504	.0216	.0081	.0027

The probability histogram is given in Figure 4.1.

4.53 (a) $F(9; 12) = .242$

(b) $f(9; 12) =$ F(9; 12) $-$ F(8; 12) $=$ (.242) $-$ (.155) $= .087$

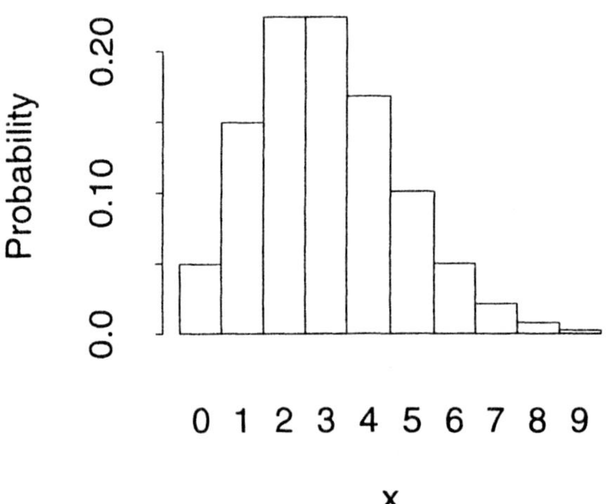

Figure 4.1: Probability Histogram for Exercise 4.51.

(c)

$$\sum_{k=3}^{12} f(k; 7.5) = F(12; 7.5) - F(2; 7.5) = (.9573) - (.0203) = .937$$

where we have interpolated, by eye, between the entries for $\lambda = 7.4$ and $\lambda = 7.6$.

4.55 (a) $n = 80$, $p = .06$, $np = 4.8$. Thus,

$$f(4; 8.4) = F(4; 4.8) - F(3; 4.8) = (.476) - (.294) = .182$$

(b) $1 - F(2; 4.8) = 1 - (.143) = .857$

(c)

$$\sum_{k=3}^{6} f(k; 4.8) = F(6; 4.8) - F(2; 4.8) = (.791) - (.143) = .648$$

4.57 $1 - F(12; 5.8) = 1 - .993 = .007$.

4.59 (a) $P(\text{at most 4 in a minute}) = F(4;\ 1.5\ ) = .981$.

(b) $P(\text{at least 3 in 2 minutes}) = 1 - F(2; 3) = 1 - (.423) = .577$.

(c) $P(\text{at most 15 in 6 minutes}) = F(15; 9) = .978$.

4.61 By definition,

$$b(1; x, p) = \binom{x}{1} p^1 (1-p)^{x-1} = xp^1(1-p)^{x-1}$$

so,

$$\frac{b(1; x, p)}{x} = p(1-p)^{x-1} = g(x; p).$$

(a) $g(12; .1) = b(1; 12, .1)/12 = (.6590 - .2824)/12 = .0314$.

(b) $g(10; .3) = b(1; 10, .3)/10 = (.1493 - .0282)/10 = .0121$.

4.63 P (fails after 1,200 times)

$$= \sum_{x=1201}^{\infty} (1-p)^{x-1} p = \frac{(1-p)^{1200}\ p}{1-(1-p)} = (1-p)^{1200}$$

where $p = .001$. Thus,

$$P(\text{fails after 1,200 times }) = (.999)^{1200} = .3010.$$

4.65 The Poisson process with $\alpha = 0.3$ applies.

(a) $\lambda = 0.3(10) = 3$.

$$f(2; 3) = \frac{3^2}{2!} e^{-3} = 0.2240.$$

(b) We calculate $1 - f(0; 3) - f(1; 3) = 1 - 4e^{-3} = .8009$, or using Table 2,

$$1 - F(1; 3) = 1 - 0.199 = 0.801.$$

(c) For either of the two 5-minute spans, the probability of exactly 1 customer is

$$f(1; 1.5) = \frac{(1.5)}{1!} e^{-1.5} = .3347$$

The two time intervals do not overlap so the counts are independent and we multiply the two probabilities

$$f(1; 1.5) \times f(1; 1.5) = (.3347) \times (.3347) = .1120$$

4.67 The Poisson process with $\alpha = 0.6$ applies.

(a) $\lambda = 0.6(2) = 1.2$.

$$f(2; 1.2) = \frac{(1.2)^2}{2!} e^{-1.2} = 0.1563.$$

(b) For the first hundred feet, the probability of exactly 1 flaw is

$$f(1; 0.6) = \frac{.06}{1} e^{-0.6} = .0511$$

The intervals do not overlap so the counts are independent and we multiply the two probabilities

$$f(1; \lambda) \times f(1; \lambda) = (.3293) \times (.3293) = .1084$$

4.69

$$\begin{aligned}
\mu &= \sum_{x=0}^{\infty} x f(x, \lambda) = \sum_{x=0}^{\infty} x \, \frac{(\lambda)^x}{x!} e^{-\lambda} = \lambda \sum_{x=1}^{\infty} \frac{(\lambda)^{x-1}}{(x-1)!} e^{-\lambda} \\
&= \lambda \sum_{x=0}^{\infty} \frac{(\lambda)^x}{x!} e^{-\lambda} \quad = \quad \lambda \\
\mu_2' &= \sum_{x=0}^{\infty} x^2 f(x, \lambda) = \sum_{x=0}^{\infty} x^2 \, \frac{(\lambda)^x}{x!} e^{-\lambda} = \sum_{x=1}^{\infty} \frac{x(\lambda)^x}{(x-1)!} e^{-\lambda} \\
&= \lambda \sum_{x=0}^{\infty} (x+1) \frac{(\lambda)^x}{x!} e^{-\lambda} \quad = \quad \lambda(\lambda + 1)
\end{aligned}$$

Thus,

$$\sigma^2 = \mu_2' - \mu^2 = \lambda(\lambda+1) - \lambda^2 = \lambda$$

4.71 The cumulative Poisson probabilities are

(a)

```
Poisson with mu = 2.73000

    x       P( X <= x )
 2.00          0.4863
 3.00          0.7075
```

(b)

```
Poisson with mu = 4.33000

    x       P( X <= x )
 2.00          0.1936
 3.00          0.3718
```

4.73 To find this probability , we use the multinomial distribution with $n = 6$, $x_1 = 2$, $x_2 = 3$, $x_3 = 1$, $p_1 = 1/4$, $p_2 = 1/2$ and $p_3 = 1/4$, Thus the probability is given by

$$\frac{6!}{2!\,3!\,1!}(\frac{1}{4})^2\,(\frac{1}{2})^3\,(\frac{1}{4})^1 = .117$$

4.75 (a) Let $\sum_{i=1}^{k} a_i = N$ and $\sum_{i=1}^{k} x_i = n$. Then the formula for the probability of getting x_i objects of the i-th kind when there are a_i objects of the i-th kind is

$$\frac{\binom{a_1}{x_1}\binom{a_2}{x_2}\cdots\binom{a_k}{x_k}}{\binom{N}{n}} = \frac{\prod_{i=1}^{k}\binom{a_i}{x_i}}{\binom{N}{n}}$$

(b) Here $a_1 = 10$, $a_2 = 7$, $a_3 = 3$, with $n = 6$, $x_1 = 3$, $x_2 = 2$, and $x_3 = 1$. Thus the probability is

$$\frac{\binom{10}{3}\binom{7}{2}\binom{3}{1}}{\binom{20}{6}} = \frac{120 \cdot 21 \cdot 3}{38,760} = .1950$$

4.77 We will use column 4 of Table 7 for this simulation. Starting in row 1, and omitting 0, 7, 8, and 9 gives:

Roll no.	1	2	3	4	5	6	7	8	9	10	11	12
Digit	6	2	5	5	1	6	3	5	6	2	6	2
Roll no.	13	14	15	16	17	18	19	20	21	22	23	24
Digit	2	4	3	3	4	5	5	3	4	2	6	1
Roll no.	25	26	27	28	28	30	31	32	33	34	35	36
Digit	3	4	2	3	4	6	5	4	1	5	2	5
Roll no.	37	38	39	40	41	42	43	44	45	46	47	48
Digit	2	4	4	6	1	5	4	1	3	5	6	4
Roll no.	49	50	51	52	53	54	55	56	57	58	59	60
Digit	6	6	6	1	2	5	1	5	1	6	4	3
Roll no.	61	62	63	64	65	66	67	68	69	70	71	72
Digit	4	1	1	5	2	5	4	5	4	4	5	5
Roll no.	73	74	75	76	77	78	79	80	81	82	83	84
Digit	4	5	2	5	4	6	3	4	2	1	1	5
Roll no.	85	86	87	88	89	90	91	92	93	94	95	96
Digit	1	5	1	6	5	6	3	2	2	3	6	1

Roll no.	97	98	99	100	101	102	103	104	105	106	107	108
Digit	2	6	2	1	2	2	6	5	5	2	1	6

Roll no.	109	110	111	112	113	114	115	116	117	118	119	120
Digit	3	1	4	2	6	3	3	1	3	6	6	1

Since the table ran out of random digits in column 4, we started over in column 8, row 1.

4.79 The random numbers are distributed in the following table:

Number of claims	Probability	Cumulative probability	Random numbers
0	.041	.041	000–040
1	.130	.171	041–170
2	.209	.380	171–379
3	.223	.603	380–602
4	.178	.781	603–780
5	.114	.895	781–894
6	.060	.955	895–954
7	.028	.983	955–982
8	.011	.994	983–993
9	.004	.998	994–997
10	.002	1.000	998-999

Using columns 5-7 of Table 7, and starting in row 1, gives:

Year	1	2	3	4	5	6	7	8	9	10
Random no.	118	243	202	654	693	785	055	418	564	679
No. of failed generators	1	2	2	4	4	5	1	3	3	4

Year	11	12	13	14	15	16	17	18	19	20
Random no.	608	095	706	919	912	295	705	920	632	449
No. of failed generators	4	1	4	6	6	2	4	6	4	3

4.81 (a) $P($ 2 or more defects $) = f(2) + f(3) = .03 + .01 = .04$.

(b) 0 is more likely since its probability $f(0) = .89$ is much larger than $f(1) = .07$.

4.83 (a) $\mu = 0 \times .07 \ + \ 1 \times .15 \ + \ 2 \times .45 \ + \ 3 \times .25 \ + \ 4 \times .08 \ = \ 2.12$.

(a) We first calculate

$$0^2 \times .07 \ + \ 1^2 \times .15 \ + \ 2^2 \times .45 \ + \ 3^2 \times .25 \ + \ 4^2 \times .08 \ = \ 5.48$$

so variance $= 5.48 - (2.12)^2 = .9856$

(c) standard deviation $= \sqrt{.9856} = .9928$ rooms

4.85 (a) Yes. $0 \leq f(i) \leq 1$, and $\sum_{i=0}^{4} f(i) \ = \ 1$.

(b) Yes. $0 \leq f(i) \leq 1$, and $\sum_{i=-1}^{1} f(i) \ = \ 1$.

(c) No. $\sum_{i=0}^{3} f(i) \ = \ 1.5 > 1$

4.87 (a)

$$b(3; 8, .2) = \binom{8}{3} (.2)^3 \, (.8)^5 = \frac{8!}{3! \; 5!} (.2)^3 \, (.8)^5 = .1468$$

(b) $B(3; 8, .2) - B(2; 8, .2) = .9437 - .7969 = .1468$

4.89 Using the hypergeometric distribution,

(a)

$$h(0;2,4,15) = \frac{\binom{4}{0}\binom{11}{2}}{\binom{15}{2}} = \frac{1 \cdot 55}{105} = .5238$$

(b)

$$h(1;2,4,15) = \frac{\binom{4}{1}\binom{11}{1}}{\binom{15}{2}} = \frac{4 \cdot 11}{105} = .4190$$

(c)

$$h(2;2,4,15) = \frac{\binom{4}{2}\binom{11}{0}}{\binom{15}{2}} = \frac{6 \cdot 1}{105} = .0571$$

4.91 (a) The variance is given by:

$$\begin{aligned} \sigma^2 &= (0-1.2)^2(.216) + (1-1.2)^2(.432) + (2-1.2)^2(.288) + (3-1.2)^2(.064) \\ &= .72 \end{aligned}$$

(b) Using the special formula for the binomial variance

$$\sigma^2 = np(1-p) = 3(.4)(.6) = .72$$

4.93 We use the approximation $b(x; n, p) \approx f(x; np)$. Thus

$$f(2; 100(.02)) = f(2; 2) = 2^2 e^{-2}/2! = .2707.$$

4.95 Since $(202 - 142)/12 = (142 - 82)/12 = 5$, we can apply Chebyshev's theorem with $k = 5$. Let X be the number of orders filled. Then,

$$P(X \leq 82 \ \text{ or } X \geq 202) = P(|X - 142| \geq 5 \cdot 12) \leq \frac{1}{25}$$

Thus,

$$P(82 < X < 202) > \frac{24}{25} = .96$$

4.97 $\lambda = .6$ for two weeks. We need

$$f(0; 6) = (.6)^0 e^{-.6}/0! = .5488.$$

4.99 (a) The random numbers are distributed in the following table:

Number of spills	Probability	Cumulative probability	Random numbers
0	.2466	.2466	0000–2465
1	.3452	.5918	2466–5917
2	.2417	.8335	5918–8334
3	.1128	.9463	8335–9462
4	.0395	.9858	9463–9857
5	.0111	.9969	9858–9968
6	.0026	.9995	9969–9994
6	.0005	1.0000	9995–9999

(b) Using columns 9-12 of the third page of Table 7 and starting from row 101, gives:

Day	1	2	3	4	5
Random no.	8353	6862	0717	2171	3763
No. of spills	3	2	0	0	1
Day	6	7	8	9	10
Random no.	1230	6120	3443	9014	4124
No. of spills	0	2	1	3	1
Day	11	12	13	14	15
Random no.	7299	0127	5056	0314	9869
No. of spills	2	0	1	0	5
Day	16	17	18	19	20
Random no.	6251	4972	1354	3695	8898
No. of spills	2	1	0	1	3
Day	21	22	23	24	25
Random no.	1516	8319	3687	6065	3123
No. of spills	0	2	1	2	1
Day	26	27	28	29	30
Random no.	4802	8030	6960	1127	7749
No. of spills	1	2	2	0	2

Chapter 5

PROBABILITY DENSITIES

5.1

$$f(x) = \begin{cases} 2e^{-2x} & \text{for } x > 0 \\ 0 & \text{elsewhere} \end{cases}$$

Since e^{-2x} is always positive, $f(x)$ is always ≥ 0.

$$\int_{-\infty}^{\infty} f(x)dx = -e^{-2x}\Big|_0^{\infty} = 1$$

Thus, $f(x)$ is a density.

5.3 The distribution function is given by

$$F(x) = \int_{-\infty}^{x} f(s)ds = x^4$$

(a) $P(X > .8) = 1 - F(.8) = .5904$

(b) $P(.2 < X < .4) = F(.4) - F(.2) = .024$

5.5

$$F(x) = \int_{-\infty}^{x} f(s)ds = \begin{cases} 0 & x < 0 \\ x^2/2 & 0 \le x \le 1 \\ 1/2 + [2s - s^2/2]|_1^x & 1 < x \le 2 \\ 1 & x > 2 \end{cases}$$

$$= \begin{cases} 0 & x < 0 \\ x^2/2 & 0 \le x \le 1 \\ 2x - x^2/2 - 1 & 1 < x \le 2 \\ 1 & x > 2 \end{cases}$$

(a) $P(X > 1.8) = 1 - F(1.8) = 1 - [2(1.8) - (1.8)^2/2 - 1] = 1 - .98 = .02$

(b) $P(.4 < X < 1.6) = F(1.6) - F(.4) = 2(1.6) - (1.6)^2/2 - 1 - (.4)^2/2 = .84$

5.7 Let X have distribution $F(x)$. Then,

(a) $P(X < 3) = F(3) = 1 - 4/9 = 5/9 = .556$

(b) $P(4 \le X \le 5) = F(5) - F(4) = 4/16 - 4/25 = .09$

5.9 (a) $P(0 \le \text{error} \le \pi/4) = \int_0^{\pi/4} \cos x dx = \sin x|_0^{\pi/4} = \sin(\pi/4) = \sqrt{2}/2$

(b) $P(\text{phase error} > \pi/3) = \int_{\pi/3}^{\pi/2} \cos x dx = \sin(\pi/2) - \sin(\pi/3) = 1 - \sqrt{3}/2$

$= .1339$

5.11 Integrating the density function by parts shows that the distribution function is given by

$$F(x) = 1 - \frac{1}{3}xe^{-x/3} - e^{-x/3}$$

Thus,

$$\begin{aligned} & P(\text{power supply will be inadequate on any given day}) \\ = \;& P(\text{consumption} \ge 12 \text{ million kwh's}) \\ = \;& 1 - F(12) = 4e^{-4} + e^{-4} = 5e^{-4} = .0916 \end{aligned}$$

5.13 The density is

$$f(x) = \begin{cases} 4x^3 & 0 < x < 1 \\ 0 & \text{elsewhere} \end{cases}$$

Thus,

$$\mu = \int_0^1 4x^4 dx = 4x^5/5\Big|_0^1 = 4/5$$

$$\mu_2' = \int_0^1 4x^5 dx = 4x^6/6\Big|_0^1 = 2/$$

and the variance is

$$\sigma^2 = \mu_2' - \mu^2 = 2/3 - (4/5)^2 = .0267$$

5.15 The density is:

$$f(x) = \begin{cases} 8x^{-3} & x > 2 \\ 0 & x \leq 2 \end{cases}$$

Thus,

$$\mu = \int_2^\infty x(8x^{-3})dx = -8x^{-1}\Big|_2^\infty = 4$$

and

$$\mu_2' = \int_2^\infty x^2(8x^{-3})dx = 8 \ln x|_2^\infty = \infty$$

Thus, σ^2 does not exist.

5.17 The density is:

$$f(x) = \begin{cases} (1/20)e^{-x/20} & x > 0 \\ 0 & x \leq 0 \end{cases}$$

Thus,

$$\mu = \frac{1}{20}\int_0^\infty xe^{-x/20} dx$$

Integrating by parts gives:

$$\begin{aligned}\mu &= -xe^{-x/20}\Big|_0^\infty + \int_0^\infty e^{-x/20}dx \\ &= 0 - 20e^{-x/20}\Big|_0^\infty \\ &= 20 \text{ (thousand miles)}\end{aligned}$$

5.19 (a) $P(\text{less than } 1.50) = F(1.50) = .9332$

(b) $P(\text{less than } -1.20) = F(-1.20) = .1151$

(c) $P(\text{greater than } 2.16) = 1 - F(2.16) = 1 - .9846 = .0154$

(d) $P(\text{greater than } -1.75) = F(1.75) = .9599$

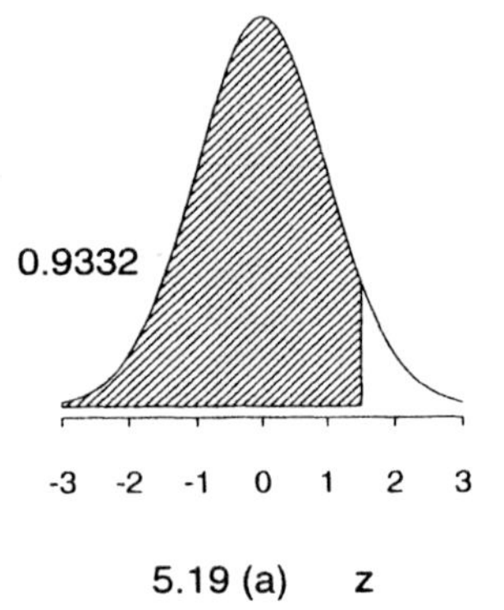

5.19 (a) z

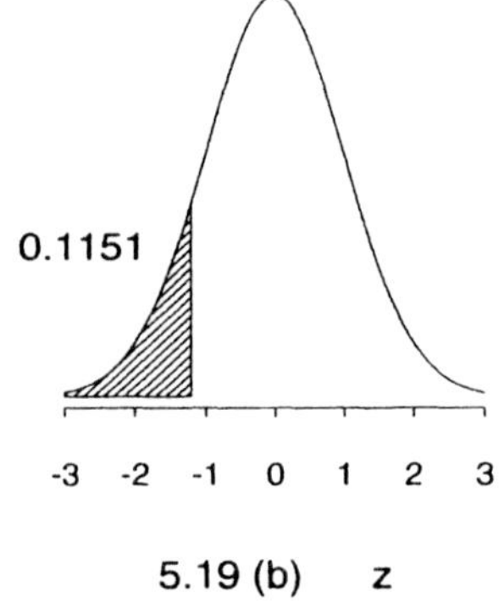

5.19 (b) z

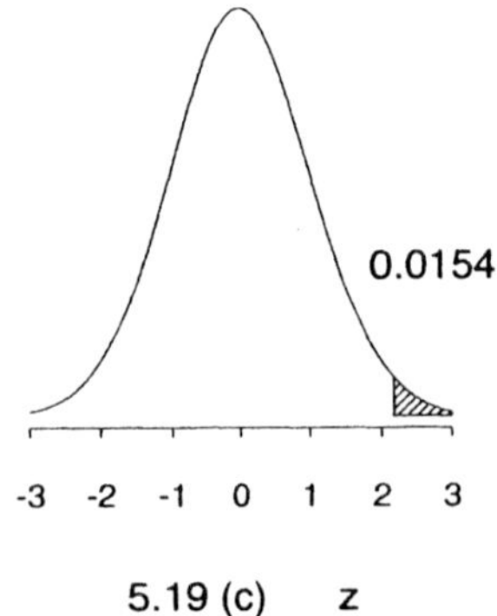

5.19 (c) z

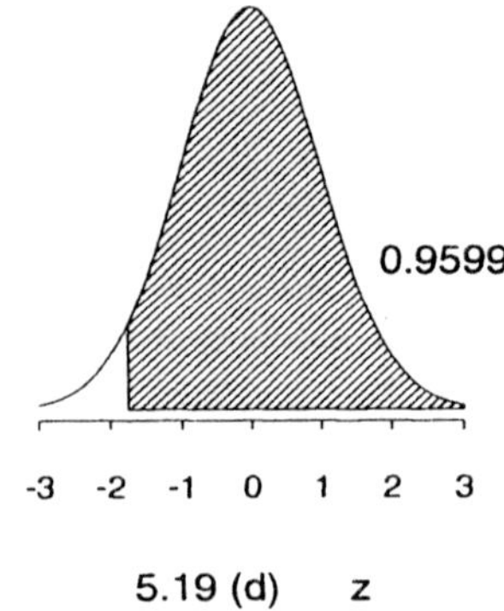

5.19 (d) z

5.21 (a) $P(Z \leq z) = F(z) = .9911$. Thus $z = 2.37$

(b) $P(Z > z) = .1093$. That is, $P(Z \leq z) = 1 - .1093$ or $F(z) = .8907$. Thus, $z = 1.23$

(c) $P(Z > z) = .6443$. That is, $F(z) = 1 - .6443 = .3557$. Using Table 3, $z = -.37$

(d) $P(Z < z) = .0217$ so z is negative. From Table 3, $z = -2.02$.

(e) $P(-z \leq Z \leq z) = .9298$. That is, $F(z) - F(-z) = .9298$, which implies that $F(z) - (1 - F(z)) = .9298$ or $F(z) = (1 + .9298)/2 = .9649$. By Table 3, $z = 1.81$.

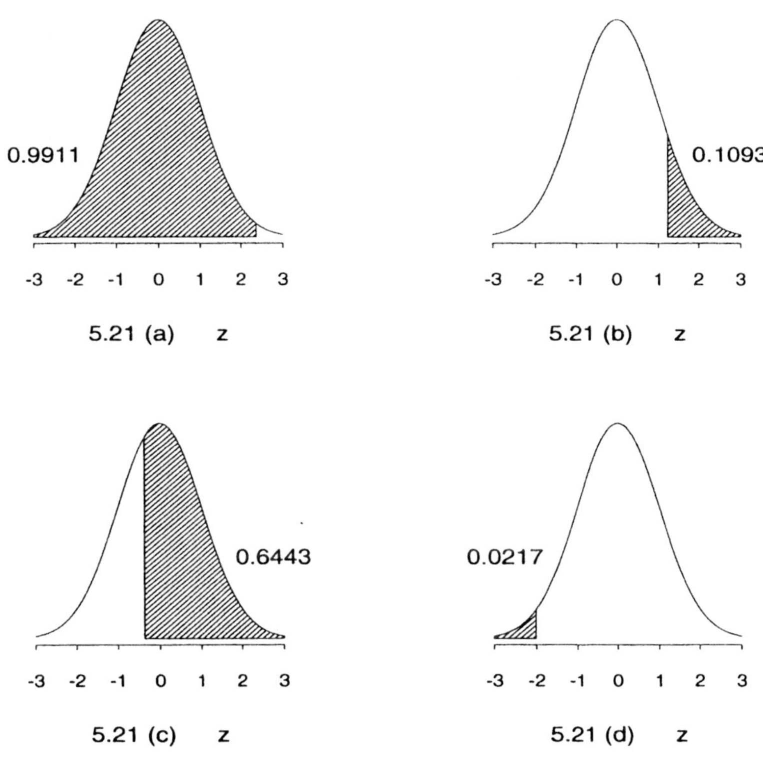

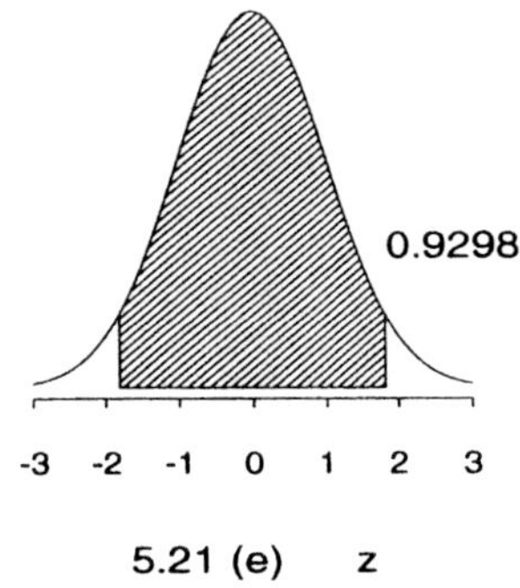

5.23 (a) $P(Z > z_{.005}) = .005$. Thus, $F(z_{.005}) = .995$ and $z = 2.575$ by linear interpolation in the Table 3.

(b) $P(Z > z_{.025}) = .025$. Thus, $F(z_{.025}) = .975$ and $z = 1.96$

5.25

$$P[X > 39.2] = .20 \quad \text{so} \quad P[\frac{X - 30}{\sigma} > \frac{9.2}{\sigma}] = .20$$

That is, $1 - F(9.2/\sigma) = .20$, and $F(9.2/\sigma) = .80$. But $F(.842) = .80$. Thus $9.2/\sigma = .842$, so $\sigma = 10.93$.

5.27 (a) We need to find $P(X > 11.5)$, where X is normally distributed with $\mu = 12.9$ and $\sigma = 2$.

$$\begin{aligned} P(X \geq 11.5) &= 1 - F((11.5 - 12.9)/2) = F((12.9 - 11.5)/2) \\ &= F(.7) = .7580 \end{aligned}$$

(b)

$$\begin{aligned} P(11 \leq X \leq 14.8) &= F((14.8 - 12.9)/2) - F((11 - 12.9)/2) \\ &= F(.95) - F(-.95) = .8289 - .1711 = .6578 \end{aligned}$$

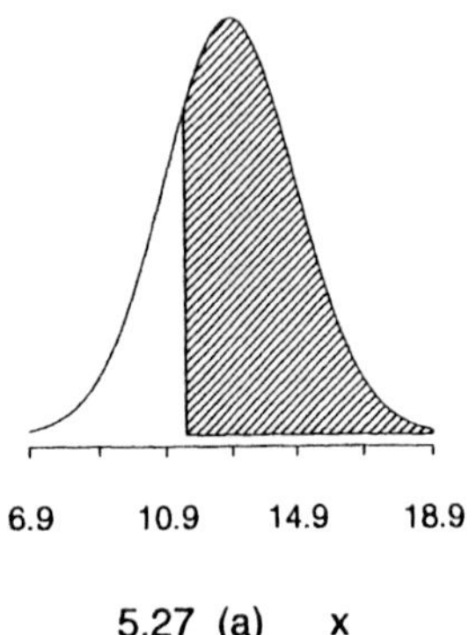

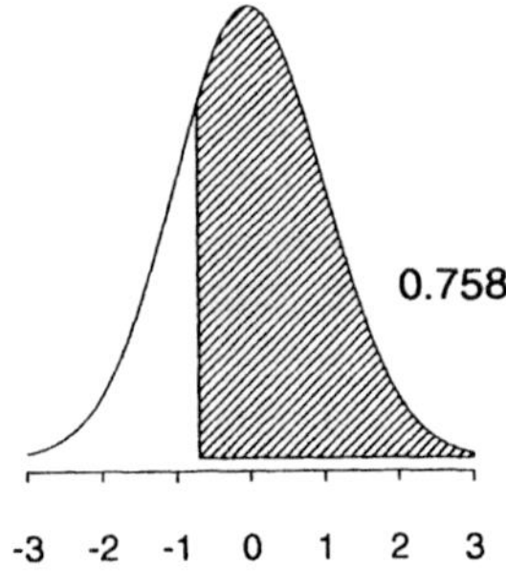

5.27 (a)

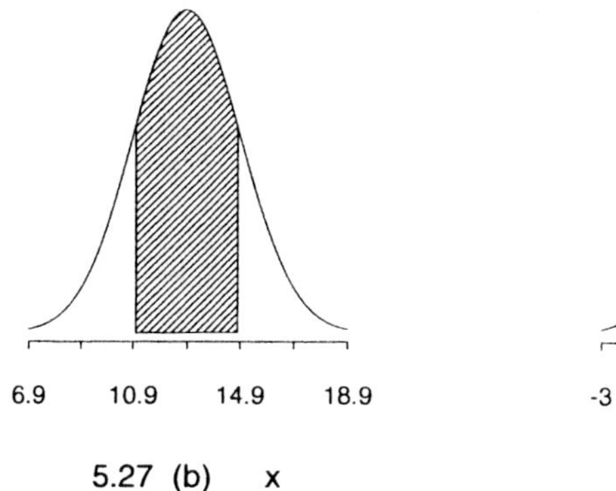

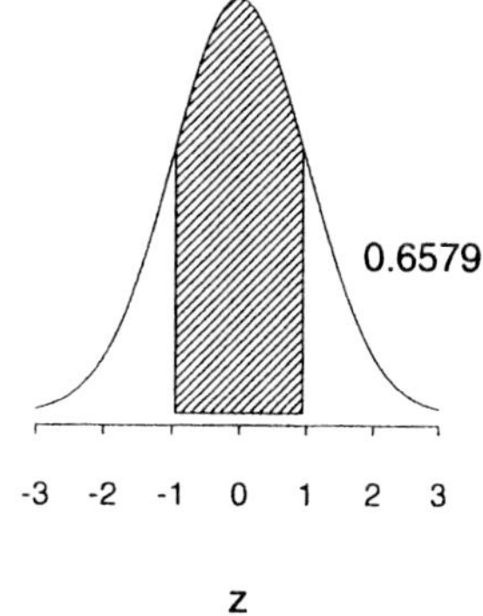

5.29 Let X be a random variable representing the developing time which is normally distributed with $\mu = 16.28$ and $\sigma = .12$.

(a) $P(16 \leq X \leq 16.5) = F((16.5 - 16.28)/.12) - F((16 - 16.28)/.12)$

$= F(1.833) - F(-2.333) = .9666 - .0098 = .9568$

(These values are determined by interpolation)

(b) $P(X \geq 16.20) = 1 - F((16.20 - 16.28)/.12) = 1 - F(-.667)$

$= F(.667) = .7476$

(c) $P(X \leq 16.35) = F((16.35 - 16.28)/.12) = F(.5833) = .7201$

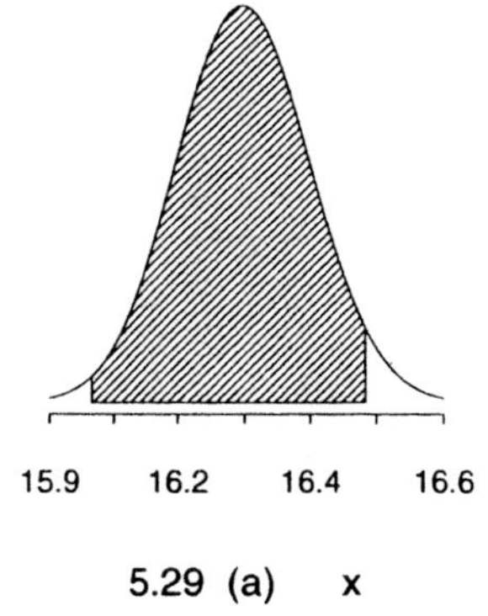

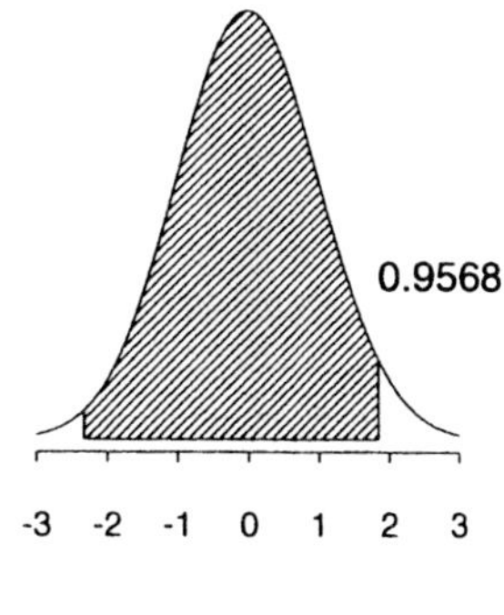

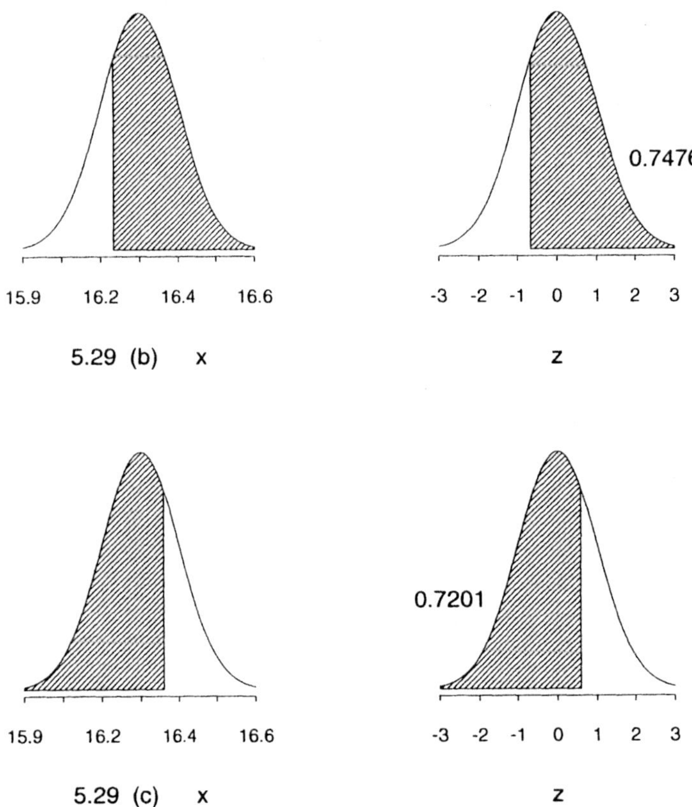

5.31 $P(.295 \leq X \leq .305) = F((.305 - .302)/.003) - F((.295 - .302)/.003)$

$= F(1) - F(-2.333) = .8413 - .0098 = .8315$

Thus, 83.15 percent will meet specifications.

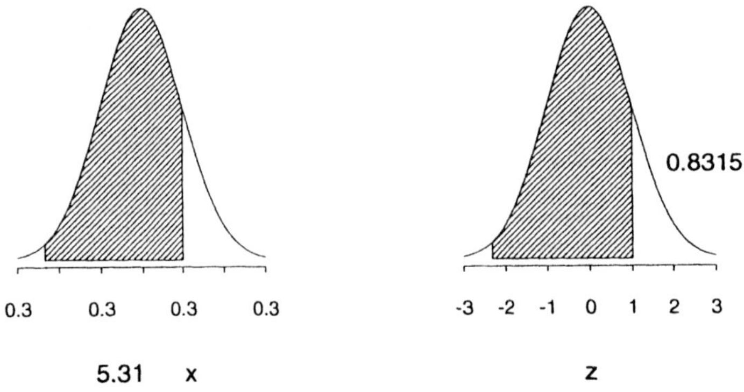

5.33 We need to find μ such that $F((3 - \mu)/.01) = .95$. Thus, from Table 3, $(3 - \mu)/.01 = 1.645$ or $\mu = 2.98355$.

5.35 If $n = 30$ and $p = .60$ then $\mu = 30(.60) = 18$ and $\sigma^2 = 30(.6)(.4) = 7.2$ or $\sigma = 2.6833$.

(a) $P(14) = F((14.5 - 18)/2.6833) - F((13.5 - 18)/2.6833)$
$= F(-1.304) - F(-1.677) = .0961 - .0468 = .0493$

(b) $P(\text{less than } 12) = F((11.5 - 18)/2.6833) = F(-2.42) = .0078$

5.37 In this case, $n = 200$, $p = .25$, $\mu = 50$, $\sigma^2 = 37.5$, $\sigma = 6.1237$. Thus,

$$\begin{aligned} P(\text{fewer than 45 fail}) &= F((44.5 - 50)/6.1237) \\ &= F(-.90) = .1841 \end{aligned}$$

5.39 Again, we will use the normal approximation to the binomial distribution. Here, $n = 40$, $p = .62$, $\mu = 24.8$, $\sigma^2 = 9.424$, $\sigma = 3.0699$. Thus,

$$F((20.5 - 24.8)/3.0699) = F(-1.40) = .0808$$

5.41 Let $f(x)$ be the standard normal density. Then $F(-z) = \int_{-\infty}^{-z} f(x)dx$. Using the change of variable, $s = -x$, and the fact that $f(x) = f(-x)$, we have

$$F(-z) = -\int_{\infty}^{z} f(-s)ds = \int_{z}^{\infty} f(s)ds = 1 - \int_{-\infty}^{z} f(s)ds = 1 - F(z)$$

5.43 We need to find

$$\begin{aligned} &\frac{1}{\sqrt{2\pi\sigma^2}} \int_{-\infty}^{\infty} (x-\mu)^2 \exp\left[-\frac{(x-\mu)^2}{2\sigma^2}\right] dx \\ &= \frac{2}{\sqrt{2\pi\sigma^2}} \int_{\mu}^{\infty} (x-\mu)^2 \exp\left[-\frac{(x-\mu)^2}{2\sigma^2}\right] dx \end{aligned}$$

since the integrand is an even function. Using the change of variable $s = (x-\mu)/\sigma$, the variance is equal to

$$\frac{2\sigma^2}{\sqrt{2\pi}}\int_0^\infty s^2\exp[-s^2/2]ds$$

Integrating by parts with $u = s$ and $dv = s\exp[-s^2/2]/(2\pi)^{1/2}ds$ shows that the variance is equal to

$$2\sigma^2\left[-s\cdot\exp(-s^2/2)/(2\pi)^{1/2}\,|_0^\infty + \int_0^\infty \exp[-s^2/2]/(2\pi)^{1/2}ds\right]$$

The first term is zero and the second is 1/2 since this is an integration of half of the standard normal density. Thus the variance is σ^2.

5.45 The uniform density is:

$$f(x) = \begin{cases} 1/(\beta-\alpha) & \alpha < x < \beta \\ 0 & \text{elsewhere} \end{cases}$$

Thus, the distribution function is

$$F(x) = \begin{cases} 1 & x \ge \beta \\ (x-\alpha)/(\beta-\alpha) & \alpha < x < \beta \\ 0 & x \le \alpha \end{cases}$$

5.47 Suppose Mr. Harris bids $(1+x)c$. Then his expected profit is:

$$\begin{aligned} &0P(\text{low bid } < (1+x)c) + xcP(\text{low bid } \ge (1+x)c) \\ &= \; xc\int_{(1+x)c}^{2c}\frac{3}{4c}ds \;=\; 3xc[2c-(1+x)c]/4c \;=\; 3c(x-x^2)/4 \end{aligned}$$

Thus, his profit is maximum when $x = 1/2$. So his bid is 3/2 times his cost. Thus, he adds 50 percent to his cost estimate.

5.49 I_0/I_i is distributed log-normal with $\alpha = 2$, $\beta^2 = .01$, $\beta = .1$. Thus,

$$P(7 \le I_0/I_i \le 7.5) \;=\; F((\ln(7.5)-2)/.1) - F((\ln(7)-2)/.1)$$

$$
\begin{aligned}
&= F(.149) - F(-.54) \\
&= .5592 - .2946 = .2646
\end{aligned}
$$

5.51 (a) $P(\text{between 3.2 and 8.4}) = F((\ln(8.4)+1)/2) - F((\ln(3.2)+1)/2)$

$= F(1.564) - F(1.0816) = .9406 - .8599 = .0807$

(b) $P(\text{greater than 5}) = 1 - F((\ln(5)+1)/2) = 1 - F(1.305) = 1 - .904 = .0960$

5.53 When $\alpha = 2$ and $\beta = 2$, $\Gamma(2) = 1$. So

$$
f(x) = \begin{cases} xe^{-x/2}/4 & x > 0 \\ 0 & x \leq 0 \end{cases}
$$

Thus,

$$
P(X < 4) = \int_0^4 f(x)dx = \frac{1}{4}\int_0^4 xe^{-x/2}dx
$$

Integrating by parts gives

$$
\begin{aligned}
-\frac{1}{2}xe^{-x/2}\,\Big|_0^4 + \frac{1}{2}\int_0^4 e^{-x/2}dx &= 2e^{-2} - e^{-x/2}\,\Big|_0^4 \\
&= 1 - 3e^{-2} = .5940
\end{aligned}
$$

5.55 (a) The probability that the supports will survive, if $\mu = 3.0$ and $\sigma^2 = .09$, is

$$
\begin{aligned}
P(\text{supports will survive}) &= 1 - F\left(\frac{\ln(33) - 3.0}{.30}\right) = 1 - F(1.655) \\
&= 1 - .9508 = .0492
\end{aligned}
$$

(b) If $\mu = 4.0$ and $\sigma^2 = .36$, then

$$
\begin{aligned}
P(\text{supports will survive}) &= 1 - F\left(\frac{\ln(33) - 4.0}{.60}\right) = 1 - F(-.84) \\
&= F(.84) = .7995
\end{aligned}
$$

5.57 We can ignore the constant in the density since it is always positive. Thus, we need to maximize $f(x) = x^{\alpha-1}e^{-x/\beta}$. Taking the derivative

$$f'(x) = (\alpha-1)x^{\alpha-2}e^{-x/\beta} - x^{\alpha-1}e^{-x/\beta}/\beta = x^{\alpha-2}e^{-x/\beta}(\alpha - 1 - x/\beta)$$

Setting the derivative equal to zero gives the solution $x = \beta(\alpha-1)$. For $\alpha > 1$, the derivative is positive for $x < \beta(\alpha-1)$ and negative for $x > \beta(\alpha-1)$. Thus, $\beta(\alpha-1)$ is a maximum. Note that $x = 0$ is a point of inflection when $\alpha > 2$. When $\alpha = 1$, $f(x) = e^{-x/\beta}$ which has a maximum in the interval $[0,\infty]$ at $x = 0$. When $0 < \alpha < 1$, the derivative does not vanish on $(0,\infty)$ and $f(x)$ is unbounded as x decreases to 0.

5.59 Since the number of breakdowns is a Poisson random variable with parameter $\lambda = .3$, the interval between breakdowns is an exponential random variable with parameter $\lambda = .3$.

(a) The probability that the interval is less than 1 week is $1 - e^{-(.3)1} = .259$ or 25.9 percent.

(b) The probability that the interval is greater than 5 weeks is $e^{-(.3)5} = .223$ or 22.3 percent.

5.61 Let N be a random variable having the Poisson distribution with parameter αt. Then $P(N = 0) = (\alpha t)^0 e^{-\alpha t}/0! = e^{-\alpha t}$. Thus, $P(\text{waiting time is} > t) = e^{-\alpha t}$ and $P(\text{waiting time is } \leq t) = 1 - e^{-\alpha t}$.

5.63 The beta density is

$$f(x) = \frac{\Gamma(\alpha+\beta)}{\Gamma(\alpha)\Gamma(\beta)}x^{\alpha-1}(1-x)^{\beta-1}$$

for $0 < x < 1$, $\alpha > 0$, and $\beta > 0$. For $\alpha = 3$ and $\beta = 3$

$$f(x) = \frac{\Gamma(6)}{\Gamma(3)\Gamma(3)} x^2(1-x)^2 = \frac{5!}{2!2!} x^2(1-x)^2 = 30(x^2 - 2x^3 + x^4)$$

Thus,

$$\begin{aligned}\int_0^1 f(x)dx &= 30\int_0^1 (x^2 - 2x^3 + x^4)dx = 30(x^3/3 - x^4/2 + x^5/5)\mid_0^1 \\ &= 30(1/3 - 1/2 + 1/5) = 1\end{aligned}$$

as required.

5.65 (a) The mean of the beta distribution is given by $\mu = \alpha/(\alpha+\beta)$. Thus, in the case where $\alpha = 1$ and $\beta = 4$, $\mu = 1/(1+4) = 1/5 = .2$

(b) When $\alpha = 1$ and $\beta = 4$, the beta density is

$$\frac{\Gamma(5)}{\Gamma(1)\Gamma(4)} x^0(1-x)^3 = \frac{4!}{0!3!}(1-x)^3 = 4(1-x)^3$$

Thus, the required probability is given by

$$4\int_{.25}^1 (1-x)^3 dx = -(1-x)^4 \mid_{.25}^1 = (.75)^4 = .3164$$

5.67 Let X be Weibull random variable with $\alpha = .1$, $\beta = .5$ representing the battery lifetime. Then the density is $f(x) = (.1)(.5)x^{-.5}e^{-.1x^{.5}}$ for $x > 0$. Thus,

$$P(X \le 100) = \int_0^{100} (.1)(.5)x^{-.5}e^{-.1x^{.5}} dx$$

Using the change of variable $y = x^{.5}$ gives:

$$P(X \le 100) = .1\int_0^{10} e^{-.1y}dy = -e^{-.1y}\mid_0^{10} = 1 - e^{-1} = .6321$$

5.69 The probability is

$$\int_{4,000}^{\infty} (.025)(.500)x^{-.5}e^{-(.025)x^{.500}}dx = \int_{\sqrt{4,000}}^{\infty} .025e^{-.025y}dy$$
$$= e^{-.025\sqrt{4,000}} = .2057$$

5.71 (a) The joint probability distribution of X_1 and X_2 is

$$f(x_1, x_2) = \frac{\binom{2}{x_1}\binom{1}{x_2}\binom{2}{2-x_1-x_2}}{\binom{5}{2}},$$

where $x_1 = 0, 1, 2,\ x_2 = 0, 1$, and $0 \le x_1 + x_2 \le 2$. The joint probability distribution $f(x_1, x_2)$ can be summarized in the following table:

		X_2		
	$f(x_1, x_2)$	0	1	Total
	0	.1	.2	.3
X_1	1	.4	.2	.6
	2	.1	0	.1
	Total	.6	.4	1

(b) Let A be the event that $X_1 + X_2 = 0$ or 1, then

$$P(A) = f(0,0) + f(0,1) + f(1,0) = .1 + .2 + .4 = .7$$

(c) By (a), the marginal distribution of X_1 is

$$f_1(0) = f(0,0) + f(0,1) = .1 + .2 = .3$$
$$f_1(1) = f(1,0) + f(1,1) = .4 + .2 = .6$$
$$f_1(2) = f(2,0) + f(2,1) = .1 + 0 = .1$$

(d) Since

$$f_2(0) = f(0,0) + f(1,0) + f(2,0) = .1 + .4 + .1 = .6,$$

the conditional probability distribution of X_1 given $X_2 = 0$ is

$$\begin{aligned} f_1(0|0) &= \frac{f(0,0)}{f_2(0)} = \frac{.1}{.6} = \frac{1}{6} \\ f_1(1|0) &= \frac{f(1,0)}{f_2(0)} = \frac{.4}{.6} = \frac{4}{6} \\ f_1(2|0) &= \frac{f(2,0)}{f_2(0)} = \frac{.1}{.6} = \frac{1}{6} \end{aligned}$$

5.73 (a) $P(X_1 < 1, X_2 < 1) = F(1,1)$

$$= \int_0^1 \int_0^1 x_1 x_2 dx_2 dx_1 = \frac{1}{2} \int_0^1 x_1 dx_1 = \frac{x_1^2}{4} \Big|_0^1 = 1/4$$

(b) The probability that the sum is less than 1 is given by:

$$\begin{aligned} \int_0^1 \int_0^{1-x_1} x_1 x_2 dx_2 dx_1 &= (1/2) \int_0^1 x_1 (1-x_1)^2 dx_1 \\ = (1/2)(x_1^4/4 - 2x_1^3/3 + x_1^2/2) \Big|_0^1 &= (1/2)(1/4 - 2/3 + 1/2) = 1/24 \end{aligned}$$

5.75 The joint distribution function is given by:

$$F(x_1, x_2) = \int_0^{x_1} \int_0^{x_2} s_1 s_2 ds_2 ds_1 = \frac{1}{2} \int_0^{x_1} x_2^2 s_1 ds_1 = x_1^2 x_2^2 / 4$$

for $0 < x_1 < 1$ and $0 < x_2 < 2$. Thus, the distribution function is

$$F(x_1, x_2) = \begin{cases} 0 & x_1 \le 0 \text{ or } x_2 \le 0 \\ x_1^2 x_2^2/4 & 0 < x_1 < 1 \text{ and } 0 < x_2 < 2 \\ x_2^2/4 & x_1 \ge 1 \text{ and } 0 < x_2 < 2 \\ x_1^2 & 0 < x_1 < 1 \text{ and } x_2 \ge 2 \\ 1 & x_1 \ge 1 \text{ and } x_2 \ge 2 \end{cases}$$

The distribution function of X_1 is

$$F_1(x_1) = \int_0^{x_1} f_1(s_1)ds_1 = \int_0^{x_1} 2s_1 ds_1 = x_1^2 \quad \text{for } 0 < x_1 < 1.$$

Thus,

$$F_1(x_1) = \begin{cases} 0 & x_1 \leq 0 \\ x_1^2 & 0 < x_1 < 1 \\ 1 & x_1 \geq 1 \end{cases}$$

Similarly,

$$F_2(x_2) = \int_0^{x_2} f_2(s_2)ds_2 = \frac{1}{2}\int_0^{x_2} s_2 ds_2 = x_2^2/4 \quad \text{for } 0 < x_2 < 2.$$

Thus,

$$F_2(x_2) = \begin{cases} 0 & x_2 \leq 0 \\ x_2^2/4 & 0 < x_2 < 2 \\ 1 & x_2 \geq 2 \end{cases}$$

It is easy to see that $F_1(x_1) \cdot F_2(x_2) = F(x_1, x_2)$. Thus, the random variables are independent.

5.77 The joint distribution function is given by

$$F(x,y) = \int_0^x \int_0^y \frac{6}{5}(u+v^2)dvdu = \frac{3x^2y}{5} + \frac{2xy^3}{5} \quad \text{for } 0 < x < 1, 0 < y < 1$$

Thus, the joint distribution is

$$F(x,y) = \begin{cases} 0 & x \leq 0 \text{ or } y \leq 0 \\ (3/5)x^2y + (2/5)xy^3 & 0 < x < 1,\ 0 < y < 1 \\ (3/5)y + (2/5)y^3 & x \geq 1,\ 0 < y < 1 \\ (3/5)x^2 + (2/5)x & 0 < x < 1,\ y \geq 1 \\ 1 & x \geq 1,\ y \geq 1 \end{cases}$$

The probability of the region in the preceeding exercise is given by

$$\begin{aligned} F(.5,.6)-F(.2,.6)-F(.5,.4)+F(.2,.4) &= .1332-.03168-.0728+.01472 \\ &= .04344 \end{aligned}$$

5.79 (a) By definition

$$f_1(x|y)=\frac{f(x,y)}{f_2(y)}=\begin{cases}(x+y^2)/(\frac{1}{2}+y^2) & \text{for } 0<y<1,\ 0<x<1 \\ 0 & \text{elsewhere.}\end{cases}$$

(b) Thus,

$$f_1(x|.5)=\frac{x+.5^2}{\frac{1}{2}+.5^2}=\begin{cases}4x/3+1/3 & \text{for } 0<x<1 \\ 0 & \text{elsewhere.}\end{cases}$$

(c) The mean is given by

$$\int_0^1 x(4x/3+1/3)dx=(4x^3/9+x^2/6)\mid_0^1=11/18$$

5.81 (a) To find k, we must integrate the density and set it equal to 1. Thus,

$$\begin{aligned} \int_0^1\int_0^2\int_0^\infty k(x+y)e^{-z}dzdydx \int_0^1\int_0^2 k(x+y)dydx \\ = \int_0^1 k(2x+2)dx=3k=1 \end{aligned}$$

Thus, $k=1/3$.

(b) $P(X<Y,\ Z>1)$

$$\begin{aligned} &= \frac{1}{3}\int_0^1\int_x^2\int_1^\infty (x+y)e^{-z}dzdydx = \frac{1}{3e}\int_0^1\int_x^2 (x+y)dydx \\ &= \frac{1}{3e}\int_0^1 (2x+2-3x^2/2)dx = 5/(6e) = .3066 \end{aligned}$$

5.83 (a) Notice that $f(x_1, x_2)$ can be factored into

$$\frac{1}{\sqrt{2\pi}\sigma} \exp\left[\frac{-1}{2\sigma^2}(x_1 - \mu_1)^2\right] \cdot \frac{1}{\sqrt{2\pi}\sigma} \exp\left[\frac{-1}{2\sigma^2}(x_2 - \mu_2)^2\right]$$

Thus, X_1 and X_2 are independent normal random variables. Thus,

$P(-8 < X_1 < 14, -9 < X_2 < 3) = P(-8 < X_1 < 14)P(-9 < X_2 < 3)$

$= (F((14-2)/10) - F((-8-2)/10))\,(F((3+2)/10) - F((-9+2)/10))$

$= (F(1.2) - F(-1))(F(.5) - F(-.7))$

$= (.8849 - .1587)(.6915 - .2420)$

$= (.7262)(.4495) = .3264$

(b) When $\mu_1 = \mu_2 = 0$ and $\sigma = 3$, the density is

$$f(x_1, x_2) = \frac{1}{2\pi\sigma} \exp\left[\frac{-1}{2\sigma^2}(x_1^2 + x_2^2)\right]$$

Let R be the region between the two circles. We need to find

$$\int_R \frac{1}{2\pi\sigma^2} \exp\left[\frac{-1}{2\sigma^2}(x_1^2 + x_2^2)\right] dx_1 dx_2$$

Changing to polar coordinates gives

$$\begin{aligned} &\int_{r=3}^{r=6} \int_0^{2\pi} \frac{1}{2\pi\sigma^2} \exp(-r^2/(2\sigma^2)) r\, d\theta\, dr \\ &= \int_3^6 \frac{1}{\sigma^2} \exp(-r^2/(2\sigma^2)) r\, dr = -\exp(-r^2/(2\sigma^2))\,\Big|_3^6 \\ &= e^{-1/2} - e^{-2} = .4712 \end{aligned}$$

5.85 The expected value of $g(X_1, X_2)$ is

$$\begin{aligned} \int_{-\infty}^{\infty} g(x_1, x_2) f(x_1, x_2) dx_1 dx_2 &= \int_0^1 \int_0^2 (x_1 + x_2) x_1 x_2 dx_2 dx_1 \\ &= \int_0^1 (2x_1^2 + 8x_1/3) dx_1 = 2 \end{aligned}$$

5.87 The area of the rectangle is $X \cdot Y$. Thus the mean of the area distribution is given by

$$\int_{L-a/2}^{L+a/2}\int_{W-b/2}^{W+b/2}\frac{xy}{ab}dydx = \int_{L-a/2}^{L+a/2}\frac{x}{a}Wdx = LW$$

The variance is given by

$$\begin{aligned}\int_{L-a/2}^{L+a/2}\int_{W-b/2}^{W+b/2}\frac{x^2y^2}{ab}dydx - (WL)^2 &= (W^2+b^2/12)(L^2+a^2/12)-(WL)^2 \\ &= \frac{1}{12}((aW)^2+(bL)^2+(ab)^2/12)\end{aligned}$$

5.89 (a) $E(X_1+X_2) = E(X_1)+E(X_2) = 1+(-1) = 0.$

(b) $Var(X_1+X_2) = Var(X_1)+Var(X_2) = 5+5 = 10.$

5.91 (a) $E(X_1+2X_2-3) = E(X_1+2X_2)-3 = E(X_1)+2E(X_2)-3 = 1+2(-2)-3$ $= -6.$

(b) $Var(X_1+2X_2-3) = Var(X_1+2X_2) = Var(X_1)+2^2Var(X_2) = 5+2^2(5)$ $= 25.$

5.93 (a) $E(X_1+X_2+\cdots+X_{20}) = E(X_1)+E(X_2)+\cdots+E(X_{20})$ $= 20(10) = 200.$

(b) $Var(X_1+X_2+\cdots+X_{20}) = Var(X_1)+Var(X_2)+\cdots+Var(X_{20})$ $= 20(3) = 60.$

5.95 (a) For any eleven observations the normal scores z_i, i=1, ..., 11, satisfy $F(z_i) = i/12$, thus using Table 3 the normal-scores are:

$$-1.38, -0.97, -0.67, -0.43, -0.21, 0, 0.21, 0.43, 0.67, 0.97, 1.38$$

(b) The observations on the times (sec.) between neutrinos are: .107, .196, .021, .283, .179, .854, .58, .19, 7.3, 1.18, 2.0. The normal-scores plot is given in Figure 5.1.

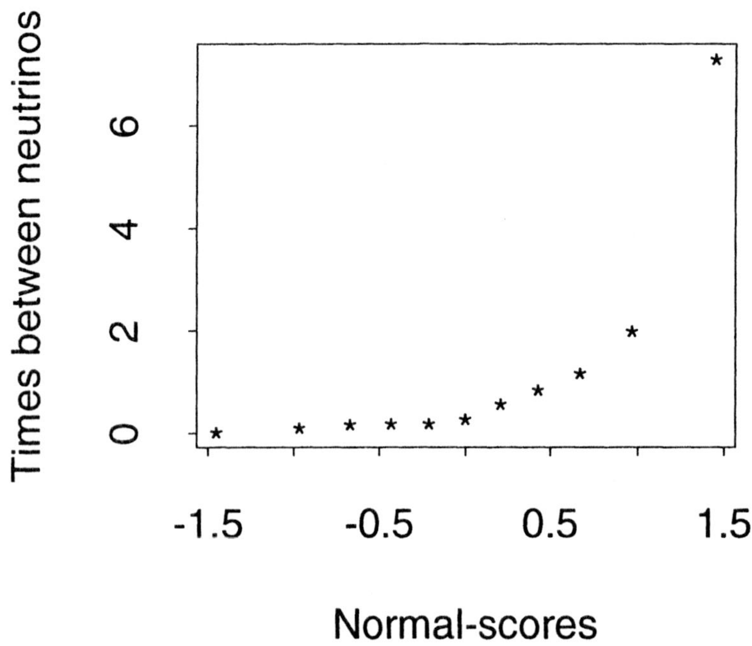

Figure 5.1: Normal-scores plot for Exercise 5.95b

5.97 (a) The normal-scores plots of the logarithmic, square root and fourth root transformations for the decay time data are given in Figures 5.2, 5.3 and 5.4, respectively.

(b) The normal-scores plots of the logarithmic, square root and fourth root transformations for the interarrival time data are given in Figures 5.5, 5.6 and 5.7, respectively.

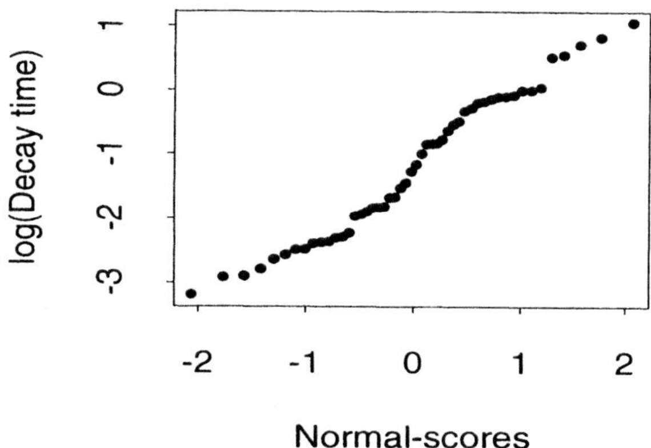

Figure 5.2: Normal-scores plot of the log(decay time) data. Exercise 5.97a

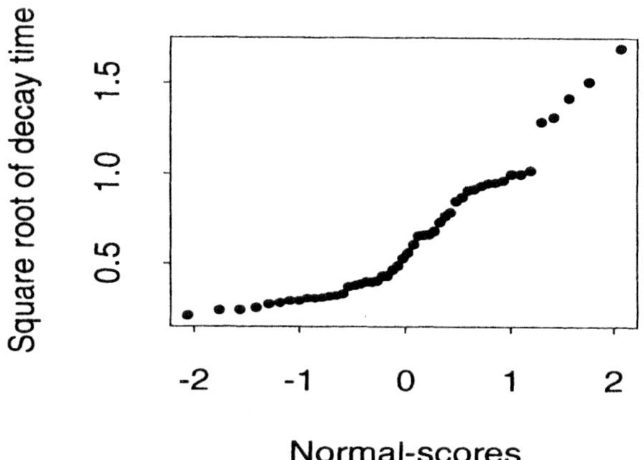

Figure 5.3: Normal-scores plot of square root of the decay time data. Exercise 5.97a

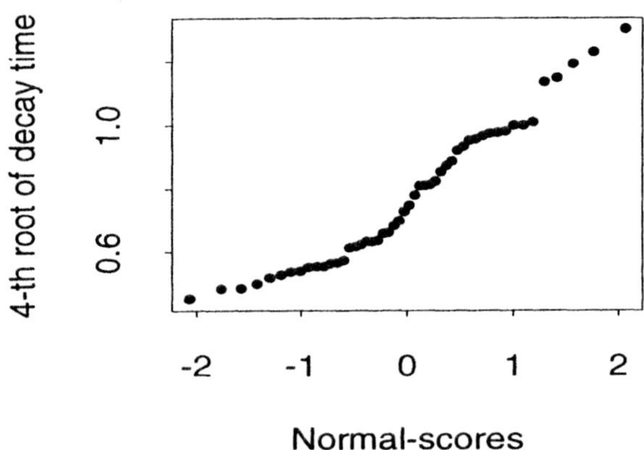

Figure 5.4: Normal-scores plot of Fourth root of the decay time data. Exercise 5.97a

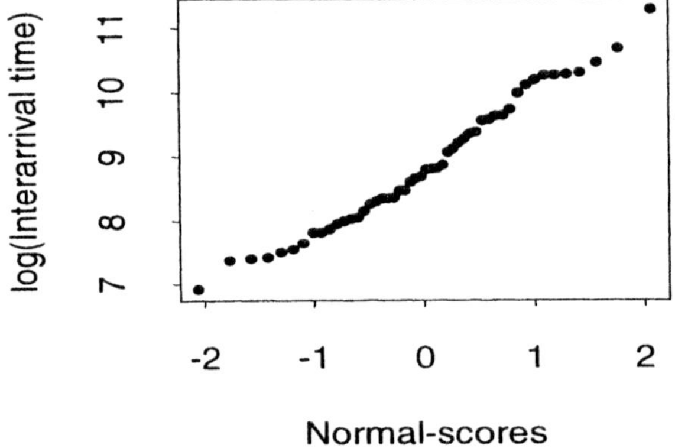

Figure 5.5: Normal-scores plot of the log(interarrival time) data. Exercise 5.97b

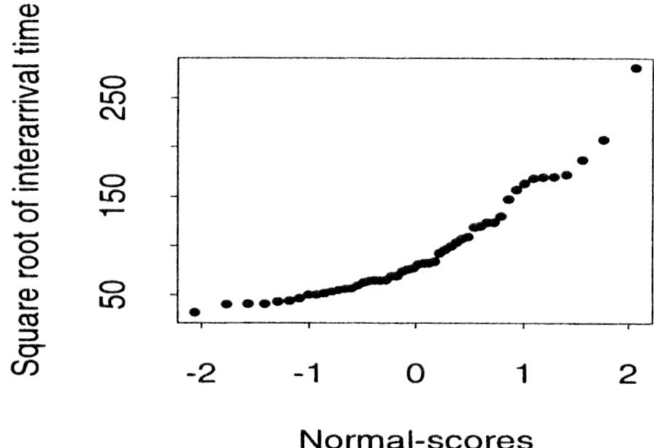

Figure 5.6: Normal-scores plot– square root of the interarrival time data. Exercise 5.97b

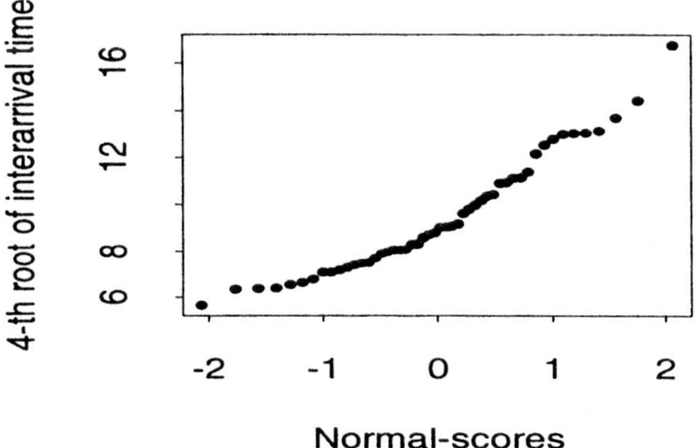

Figure 5.7: Normal-scores plot – fourth root of the interarrival time data. Exercise 5.97b

5.99 The time it takes a person to learn how to operate a certain machine is normal random variable with mean $\mu = 5.8$ and standard deviation $\sigma = 1.2$ and it takes two persons to operate the machine. Using Minitab , we generate two columns of normal variates having mean 5.8 and standard deviation 1.2. Each row represents a pair of workers.

ROW	C1	C2
1	6.16327	5.71118
2	5.37122	5.15066
3	5.21900	6.44433
4	4.73400	6.35715

The simulated times it takes four pairs to learn how to operate the machine are 6.16327, 5.37122, 6.44433 and 6.35715.

5.101 (a) The density is

$$f(x) = \begin{cases} .3e^{-.3x} & x > 0 \\ 0 & \text{elsewhere} \end{cases}$$

Thus, the corresponding distribution function is

$$F(x) = \begin{cases} \int_0^x .3e^{-.3s}ds = -e^{-.3s} \mid_0^x = 1 - e^{-.3x} & \text{for } x > 0 \\ 0 & \text{elsewhere} \end{cases}$$

(b) We solve $u = F(x)$ for x. Since $u = F(x) = 1 - e^{-.3x}$, so $e^{-.3x} = 1 - u$ or $-.3x = \ln(1-u)$. The solution is then $x = -\ln(1-u)/.3$

5.103 Let Z_1, Z_2 be independent standard normal random variables. Under a change of polar coordinates, $Z_1 = R\cos(\Theta)$, $Z_2 = R\sin(\Theta)$, the joint density of R and Θ is

$$f(r,\theta) = \begin{cases} re^{-r^2/2}\frac{1}{2\pi} & 0 < \theta < 2\pi, r > 0 \\ 0 & \text{elsewhere} \end{cases}$$

(a) The marginal distribution of Θ is

$$f_2(\theta) = \int_0^\infty f(r,\theta)dr = \frac{1}{2\pi}\int_0^\infty re^{-r^2/2}dr$$

$$= \frac{1}{2\pi}(-e^{-r^2/2}) \mid_0^\infty = \frac{1}{2\pi}.$$

Hence Θ has uniform distribution on $(0,\ 2\pi)$. The marginal distribution of R is

$$f_1(r) = \frac{1}{2\pi}\int_0^{2\pi} re^{-r^2/2}d\theta = \frac{1}{2\pi}re^{-r^2/2}\theta \mid_0^{2\pi} = re^{-r^2/2} \quad \text{for } r > 0$$

Thus, R has Weibull distribution with $\alpha = 1/2$ and $\beta = 2$. Since

$$f_1(r)f_2(\theta) = f(r,\theta)$$

R and Θ are independent.

(b) Let $U_1 = \Theta/2\pi$ and $U_2 = 1 - e^{-R^2/2}$, then U_1 and U_2 are independent since R and Θ are independent. The distribution function of U_1 is

$$F_1(u_1) = P(U_1 < u_1) = P(\Theta < 2\pi u_1) = \frac{2\pi u_1}{2\pi} = u_1 \quad \text{for } 0 < u_1 < 1$$

Thus, U_1 has uniform distribution on $(0,\ 1)$. The distribution function of U_2 is

$$\begin{aligned} F_2(u_2) &= P(U_2 < u_2) = P(1 - e^{-R^2/2} < u_2) \\ &= P(R^2 < -2\ln(1-u_2)) = P(R < (-2\ln(1-u_2))^{1/2}) \\ &= -e^{-r^2/2} \mid_0^{(-2\ln(1-u_2))^{1/2}} = u_2 \end{aligned}$$

Thus, U_2 has a uniform distribution on $(0,\ 1)$.

(c) Since $1 - U_2 = e^{-R^2/2}$ and $U_1 = \Theta/2\pi$, we have

$$Z_1 = R\cos\Theta = \sqrt{-2\ln e^{-R^2/2}}\cos\Theta = \sqrt{-2\ln(1-U_2)}\cos(2\pi U_1)$$

$$Z_2 = R\sin\Theta = \sqrt{-2\ln e^{-R^2/2}}\sin\Theta = \sqrt{-2\ln(1-U_2)}\sin(2\pi U_1)$$

Note that $1 - U_2$ also has a uniform distribution on (0, 1). Hence $\ln U_2$ can be used in place of $\ln(1 - U_2)$ in above equations, when we use independent uniform random variables U_1 and U_2 to generate independent standard normal variables Z_1 and Z_2. This completes the proof.

5.105 (a) Histogram of the time of the 100 first failures is given in Figure 5.8

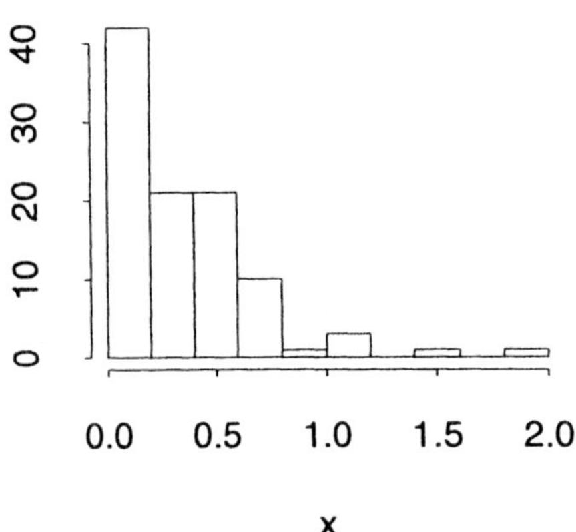

Figure 5.8: Histogram of the 100 first failure times. Exercise 5.105a.

(b) Histogram of the time of the 100 fifth failures is given in Figure 5.9. Comparing with the histogram in Part a), is shifted towards larger values.

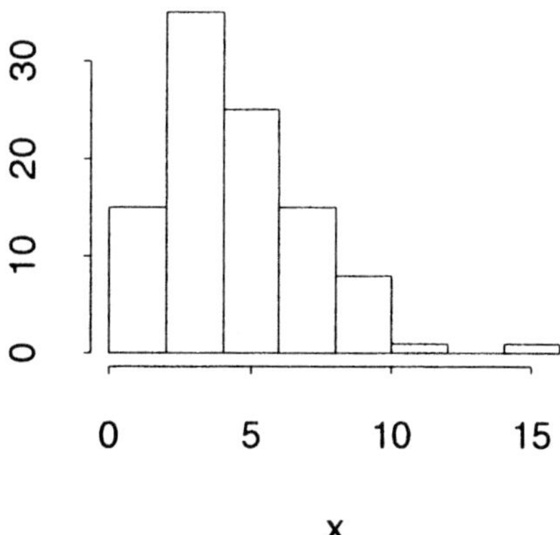

Figure 5.9: Histogram of the 100 fifth failure times. Exercise 5.105b.

5.107 (a) The histogram of the 400 learning times for the pairs of operators is given in Figure 5.10

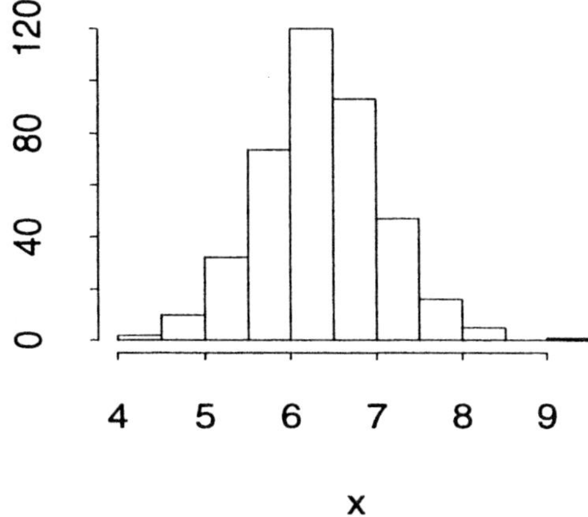

Figure 5.10: Histogram for Exercise 5.107a.

(b) The histogram of the 100 values representing the time to train four pairs of operators is given in Figure 5.11.

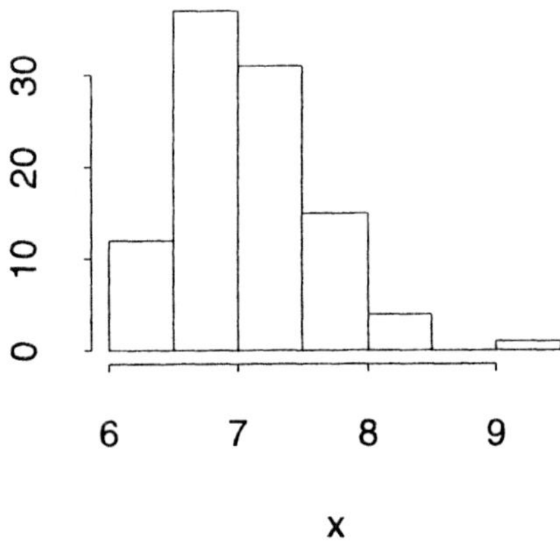

Figure 5.11: Histogram for Exercise 107b.

5.109 The distribution function is given by:

$$F(x) = \int_{-\infty}^{x} f(s)ds = \int_{0}^{x} \frac{3}{2}(1 - s^2)ds = \frac{3}{2}(x - x^3/3)$$

Thus,

(a) $P(X < .3) = F(.3) = (3/2)(.3 - .3^3/3) = .4365$

(b) $P(.4 < X < .6) = F(.6) - F(.4) = (3/2)(.6 - .6^3/3) - (3/2)(.4 - .4^3/3)$
$= .792 - .568 = .224.$

5.111

$$\mu = \int_0^1 xf(x)dx = \frac{3}{2}\int_0^1 x(1 - x^2)dx = \frac{3}{2}\left(\frac{x^2}{2} - \frac{x^4}{4}\right)\Big|_0^1 = \frac{3}{8}$$

Next,

$$\mu_2' = \frac{3}{2}\int_0^1 x^2(1 - x^2)dx = \frac{3}{2}\left(\frac{x^3}{3} - \frac{x^5}{5}\right)\Big|_0^1 = .2$$

so the variance is equal to

$$\sigma^2 = \mu_2' - \mu^2 = .2 - \left(\frac{3}{8}\right)^2 = .0594$$

5.113 Let X be a normal random variable with $\mu = 4.76$ and $\sigma = .04$

(a) $P(X < 4.66) = F((4.66 - 4.76)/.04) = F(-2.5) = .0062$

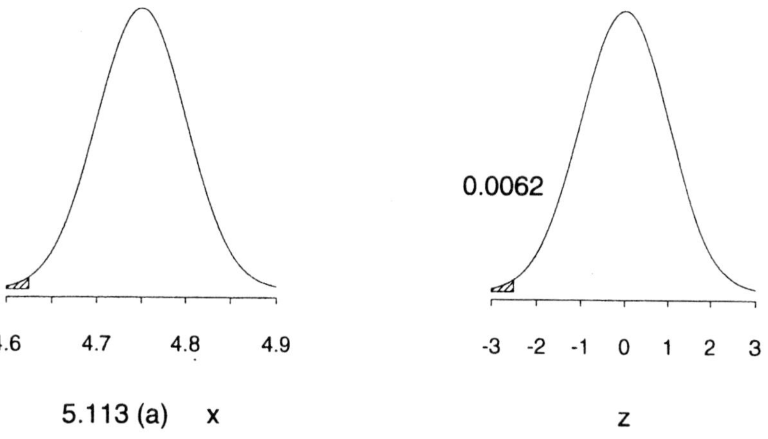

5.113 (a)

(b) $P(X > 4.8) = 1 - F((4.8 - 4.76)/.04) = 1 - F(1) = 1 - .8413 = .1587$

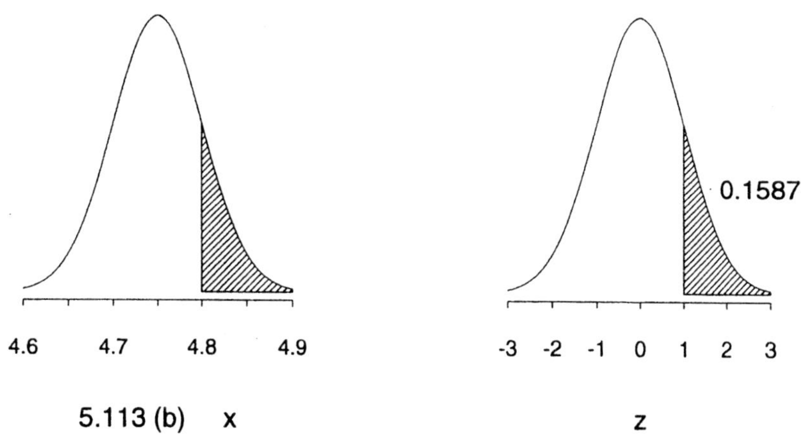

5.113 (b)

(c) $P(4.7 < X < 4.82) = F(4.82 - 4.76)/.04) - F((4.7 - 4.76)/.04)$

$= F(1.5) - F(-1.5) = .9332 - .0668 = .8664$

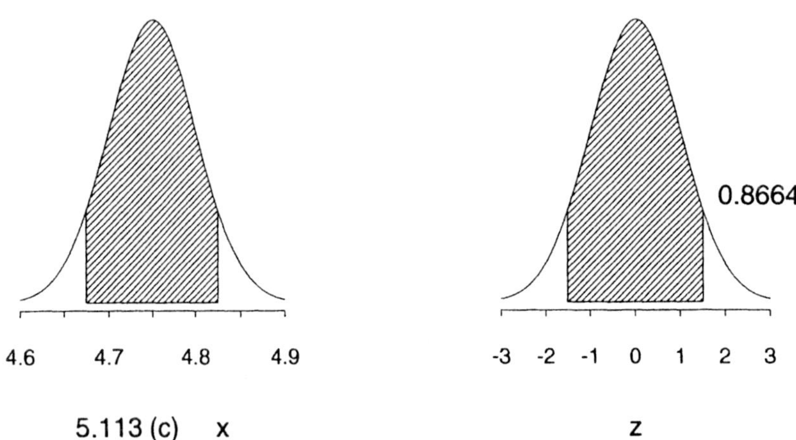

5.115 $Q_1 = -\sigma z_{.25} + \mu = -27(.675) + 102 = 83.775$

$Q_2 = \sigma z_{.5} + \mu = \mu = 102$

$Q_3 = \sigma z_{.25} + \mu = 27(.675) + 102 = 120.225$

5.117 The density function is

$$f(x) = \begin{cases} .25e^{-.25x} & x > 0 \\ 0 & \text{elsewhere} \end{cases}$$

(a) $P(\text{time to observe a particle is more than 200 microseconds})$

$= -e^{-.25x} \,|_{.2}^{\infty} = e^{-.05} = .951$

(b) $P(\text{time to observe a particle is less than 10 microseconds})$

$= 1 - e^{-.0025} = 1 - .9975 = .0025$

5.119 The normal-scores plot of the velocity of light data is given in Figure 5.12.

5.121 Let X be the strength of a support beam, having the Weibull distribution with

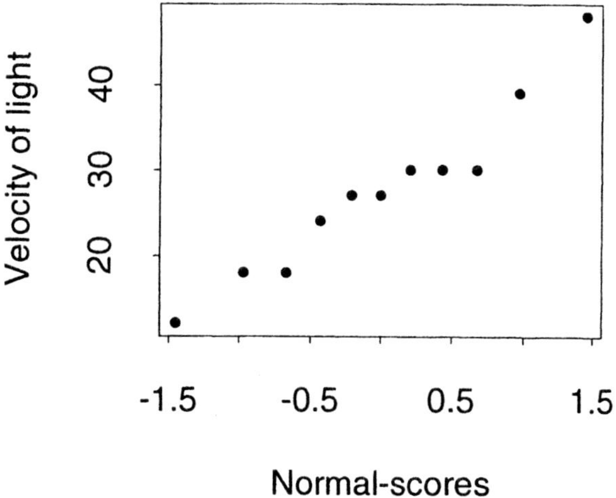

Figure 5.12: Normal-scores plot of the velocity of the light data. Exercise 5.119.

$\alpha = .02$ and $\beta = 3.0$.

$$P(X > 4.5) = \int_{4.5}^{\infty} (.02)3x^2 e^{-.02x^3} dx$$

Using the change of variable $u = x^3$, we have

$$P(X > 4.5) = \int_{(4.5)^3}^{\infty} .02e^{-.02u} du = e^{-.02(4.5)^3} = .1616$$

5.123 (a) The independent random variables X_1 and X_2 have the same probability distribution $b(x; 2, .7)$. Hence the joint probability distribution of X_1 and X_2 is

$$\begin{aligned} f(x_1, x_2) &= b(x_1; 2, .7) \cdot b(x_2; 2, .7) \\ &= \begin{pmatrix} 2 \\ x_1 \end{pmatrix} .7^{x_1} .3^{2-x_1} \cdot \begin{pmatrix} 2 \\ x_2 \end{pmatrix} .7^{x_2} .3^{2-x_2} \end{aligned}$$

$$= \binom{2}{x_1}\binom{2}{x_2}.7^{x_1+x_2}.3^{4-x_1-x_2}$$

where $x_1 = 0, 1, 2,$ and $x_2 = 0, 1, 2.$

(b)

$$\begin{aligned} P(X_1 < X_2) &= f(0,1) + f(0,2) + f(1,2) \\ &= \binom{2}{0}\binom{2}{1}.7^1.3^3 + \binom{2}{0}\binom{2}{2}.7^2.3^2 \\ &\quad + \binom{2}{1}\binom{2}{2}.7^3.3^1 \\ &= .0378 + .0441 + .2058 = .2877 \end{aligned}$$

5.125 (a) $E(X_1 + X_2 + \cdots + X_{30}) = E(X_1) + E(X_2) + \cdots + E(X_{30})$

$= 30(-5) = -150.$

(b) $Var(X_1 + X_2 + \cdots + X_{30}) = Var(X_1) + Var(X_2) + \cdots + Var(X_{30})$

$= 30(2) = 60.$

Chapter 6

SAMPLING DISTRIBUTIONS

6.1 Only if the pieces are put on the assembly line such that every 20-th piece is random with respect to the characteristic being measured.

If twenty molds are dumping their contents, in sequential order, onto the assembly line then the sample would consist of output from a single mold . It would not be random.

6.3 (a) A typical member of the population does not vacation on a luxury cruise. The sample taken would be biased.

(b) This sample will very likely be biased. Those will high incomes will tend to respond while those with low incomes will tend to not respond.

(c) Everyone feels that unfair things should be stopped. The way the question is phrased biases the responses.

6.5 (a) The number of samples (given that order does not matter) is

$$\binom{6}{2} = \frac{6 \cdot 5}{2 \cdot 1} = 15.$$

(b) The number of samples (given that order does not matter) is

$$\binom{25}{2} = \frac{25 \cdot 24}{2 \cdot 1} = 300.$$

6.7 (a) The probability of each of the numbers is given in the table:

Number	−4	−3	−2	−1	0	1	2	3	4
Probability	1/6	2/15	1/10	1/15	1/15	1/15	1/10	2/15	1/6

The mean of the distribution is

$$\frac{1}{6}(-4) + \frac{2}{15}(-3) + \cdots + \frac{2}{15}(3) + \frac{1}{6}(4)$$
$$= \frac{1}{6}(4-4) + \frac{2}{15}(3-3) + \frac{1}{10}(2-2) + \frac{1}{15}(1-1) + \frac{1}{15}(0) = 0.$$

The variance is

$$\frac{2}{6}(16) + \frac{4}{15}(9) + \frac{2}{10}(4) + \frac{2}{15}(1) + \frac{1}{15}(0) = 8.667.$$

(b) The 50 samples are shown in Table 6.1 along with their means.

Table 6.1. 50 Samples of Size 10 Taken Without Replacement

obs.	obs.	obs.	obs.	obs.	obs.	obs.	obs.	obs.	obs.	mean
1	3	-1	4	4	4	-4	-4	-4	-2	0.1
-2	3	3	-2	2	-4	4	-4	-3	-4	-0.7
4	4	-2	-3	-3	2	1	3	4	0	1.0
3	3	-3	4	2	1	-2	0	4	0	1.2
4	0	-1	-2	3	1	2	4	-1	1	1.1
1	-2	-3	-4	2	4	-2	3	4	-4	-0.1
-1	2	1	-1	-4	-3	0	-4	3	1	-0.6
-4	-2	2	1	-4	-3	3	2	-2	1	-0.6
1	-3	2	-4	2	4	3	-4	-3	-2	-0.4
1	4	4	4	-4	3	-1	-2	2	-4	0.7
4	0	-1	1	-3	1	4	-3	4	2	0.9
-4	4	2	-4	2	-2	-2	-1	4	3	0.2
2	-1	-2	4	-4	0	-4	1	3	-2	-0.3
-4	-1	-3	4	0	-4	-3	1	2	4	-0.4
-2	4	4	-1	-4	1	4	1	-4	-3	0.0
-3	0	-3	-2	0	-4	4	-1	1	-2	-1.0
4	-4	3	4	-4	-1	3	2	0	1	0.8
3	3	-3	4	-3	1	3	-3	-4	-1	0.0
3	-4	-4	4	4	3	4	-3	2	-2	0.7
4	0	2	-3	-3	-3	-2	3	-4	-4	-1.0
2	-1	4	-3	2	3	4	4	-4	4	1.5
3	2	3	4	4	4	3	1	2	-2	2.4
4	-3	0	-1	-3	-3	0	-4	4	-4	-1.0
-3	3	-4	3	0	1	-4	-4	-3	-4	-1.5
-2	-4	-2	-3	4	4	3	-3	1	-3	-0.5
3	-1	-2	-3	1	-4	-2	3	4	2	0.1
4	2	-4	3	-2	2	-2	1	4	3	1.1
0	-2	4	-2	0	-1	2	3	-3	-4	-0.3

Table 6.1.(continued)

obs.	obs.	obs.	obs.	obs.	obs.	obs.	obs.	obs.	obs.	mean
1	-4	4	4	-2	-3	-4	-4	0	3	-0.5
2	4	3	-2	3	1	0	0	-4	-3	0.4
4	0	2	-1	1	-4	-4	-4	-3	4	-0.5
4	-3	-1	-3	-3	0	4	4	3	1	0.6
-4	-1	0	-3	4	-3	2	1	3	4	0.3
0	4	-4	3	-2	2	0	4	2	-4	0.5
3	2	-2	4	-4	4	3	4	0	-3	1.1
-4	2	0	-2	-3	2	-1	-2	-4	-2	-1.4
-3	-1	-2	4	2	4	0	1	2	2	0.9
2	-4	-2	-3	-1	3	1	-2	-4	-3	-1.3
-4	3	1	0	-1	0	-3	-3	2	4	-0.1
2	2	4	-4	-4	4	4	-3	-3	-2	0.0
-4	1	4	4	3	2	-2	-3	4	-3	0.6
0	-4	-4	0	-4	1	4	-3	1	4	-0.5
-4	4	-3	1	4	-1	-4	-3	-4	-2	-1.2
4	-3	-3	4	4	-3	1	2	3	-1	0.8
-2	3	2	-4	-1	3	-4	-1	-3	1	-0.6
4	2	4	-3	-4	-4	-1	-3	-4	-2	-1.1
-3	-1	1	2	-3	0	-4	-2	4	2	-0.4
0	-1	4	-4	2	-2	3	-3	-2	3	0.0
-4	2	1	-2	-2	-1	-3	-4	3	-3	-1.3
-4	4	4	4	-3	2	3	3	3	4	2.0

(c) The mean of the 50 sample means is .034. The sample variance of the 50 sample means is .8096.

(d) According to Theorem 6.1, the distribution of the means has mean 0 and variance

$$\frac{\sigma^2}{n} \cdot \frac{N-n}{N-1} = \frac{8.667}{10} \cdot \frac{20}{29} = .5977.$$

The sample values in part (c) compare well with these theoretical values.

6.9 A table of each outcome and the mean follows:

Outcome	Mean	Outcome	Mean
1, 1	1.0	3, 1	2.0
1, 2	1.5	3, 2	2.5
1, 3	2.0	3, 3	3.0
1, 4	2.5	3, 4	3.5
2, 1	1.5	4, 1	2.5
2, 2	2.0	4, 2	3.0
2, 3	2.5	4, 3	3.5
2, 4	3.0	4, 4	4.0

Thus , the distribution of these values is:

Value	No. of Ways Obtained	Probability
1.0	1	.0625
1.5	2	.1250
2.0	3	.1875
2.5	4	.2500
3.0	3	.1875
3.5	2	.1250
4.0	1	.0625

where the probability is (*no. ways*)$\times(.25)^2$. Consequently, the distribution of $\bar{X}$ has mean

$$\mu_{\bar{X}} = 1(.0625) + 1.5(.1250) + \cdots + 3.5(.1250) + 4(.0625) = 2.5$$

which may be obtained directly from the symmetry of the distribution. The distribution of $\bar{X}$ has variance

$$\sigma^2_{\bar{X}} = (1 - 2.5)^2(.0625) + (1.5 - 2.5)^2(.1250) + \cdots$$

$$+ (3.5 - 2.5)^2(.1250) + (4 - 2.5)^2(.0625)$$

$$= .625.$$

Now the mean of the original distribution is also 2.5 and the variance is 1.25. Thus, Theorem 6.1 yields 2.5 and $1.25/2 = .625$ as the mean and the variance of the distribution of the sample mean of two observations. These agree exactly as they must.

6.11 The variance of the sample mean $\bar{X}$, based on a sample of size n, is σ^2/n. Thus the standard deviation, or standard error of the mean is $\sigma/\sqrt{n}$.

(a) The standard deviation for a sample of size 50 is $\sigma/\sqrt{50}$. The standard deviation for a sample is size 200 is $\sigma/\sqrt{200}$. That is, the ratio of standard errors is

$$\frac{\sigma/\sqrt{200}}{\sigma/\sqrt{50}} = \frac{\sqrt{50}}{\sqrt{200}} = \frac{1}{2},$$

so the standard error is halved.

(b) The ratio of standard errors is

$$\frac{\sigma/\sqrt{900}}{\sigma/\sqrt{400}} = \frac{\sqrt{400}}{\sqrt{900}} = \frac{2}{3},$$

so the standard error for a sample size 900 is 2/3rd's that for sample size 400.

(c) The standard error for a sample size 25 is

$$\frac{\sqrt{225}}{\sqrt{25}} = 3$$

times as large as that for sample size 225.

(d) The standard error for a sample size 40 is

$$\frac{\sqrt{640}}{\sqrt{40}} = 4$$

times as large as that for sample size 640.

6.13 We need to find $P(|\bar{X} - \mu| < .6745 \cdot \sigma/\sqrt{n})$. Since the standard deviation of the mean is $\sigma/\sqrt{n}$, the standardized variable $(\bar{X} - \mu)/(\sigma/\sqrt{n})$ is approximately a normal random variable for large n (central limit theorem). Thus , we need to find:

$$P(|\frac{\bar{X} - \mu}{\sigma/\sqrt{n}}| < .6745).$$

Now, interpolating in Table 3 gives

$$P(\frac{\bar{X} - \mu}{\sigma/\sqrt{n}} < .6745) = .75.$$

Thus,

$$P(\frac{\bar{X} - \mu}{\sigma/\sqrt{n}} \leq -.6745) = P(\frac{\bar{X} - \mu}{\sigma/\sqrt{n}} \geq .6745) = .25$$

so

$$P(|\frac{\bar{X} - \mu}{\sigma/\sqrt{n}}| < .6745) = .75 - .25 = .50.$$

The probability that the mean of a random sample of size n , from a population with standard deviation σ, will differ from μ by less than $(.6745)(\sigma/\sqrt{n})$ is approximately .5 for sufficiently large n.

6.15 Here $\sigma/\sqrt{n} = 16/10 = 1.6$ and

$$P(175 \leq \bar{X} \leq 178) = P(-1 \leq \bar{X} - 176 \leq 2) = P(-.625 \leq \frac{\bar{X} - 176}{16/10} \leq 1.25).$$

Since $n = 100$ is large , the central limit theorem yields the approximation

$$\begin{aligned} P(-.625 \leq \frac{\bar{X} - 176}{16/10} \leq 1.25) &\approx F(1.25) - F(-.625) \\ &= .8944 - .266) = .628. \end{aligned}$$

6.17 We need to find

$$
\begin{aligned}
P(\sum_{i=1}^{36} X_i > 6{,}000) &= P(\bar{X} > 166.67) = P(\bar{X} - 163 > 3.67) \\
&= P(\frac{\bar{X} - 163}{18/6} > 1.222).
\end{aligned}
$$

Since $n = 36$ is relatively large, we use the central limit theorem to approximate this probability by

$$1 - F(1.222) = .111.$$

6.19 Let $\mu_X = E(X)$ be the expected value of X. First we will show that $E(X+Y) = E(X) + E(Y)$. Let $f(x, y)$ be the joint density function of X and Y. Then, if X takes on discrete values x_i and Y takes on discrete values y_j,

$$
\begin{aligned}
E(X+Y) &= \sum_{i=0}^{\infty}\sum_{j=0}^{\infty}(x_i + y_j) f(x_i, y_j) \\
&= \sum_{i=0}^{\infty} x_i \sum_{j=0}^{\infty} f(x_i, y_j) + \sum_{j=0}^{\infty} y_j \sum_{i=0}^{\infty} f(x_i, y_j).
\end{aligned}
$$

But, $\sum_{j=0}^{\infty} f(x_i, y_j) = f_1(x)$, the marginal density of X. Similarly for Y. Thus

$$E(X+Y) = \sum_{i=0}^{\infty} x_i f_1(x_i) + \sum_{j=0}^{\infty} y_j f_2(y_j) = E(X) + E(Y)$$

by definition. By induction

$$E(\sum_{i=1}^{n} X_i) = \sum_{i=1}^{n} E(X_i).$$

Thus,

$$
\begin{aligned}
\mu_{\bar{X}} &= E(\frac{\sum_{i=1}^{n} X_i}{n}) \\
&= \frac{1}{n} E(\sum_{i=1}^{n} X_i) = \frac{1}{n}\sum_{i=1}^{n} E(X_i).
\end{aligned}
$$

But, each $E(X_i) = \mu$ so $\mu_{\bar{X}} = \mu$.

6.21 The mean of the data is $\bar{x} = 23$ and the sample standard deviation is 6.39. Thus, if the data is from a normal population with $\mu = 20$, the statistic

$$t = \frac{\bar{x} - \mu}{s/\sqrt{n}} = \frac{23 - 20}{6.39/\sqrt{6}} = 1.15$$

is the value of a t random variable with 5 degrees of freedom. The entry in Table 4 for $\alpha = 10$ and $\nu = 5$ is 1.476. Before the data are observed, we know that

$$P(\frac{\bar{X} - \mu}{S/\sqrt{n}} > 1.15) > .10.$$

Thus, the data does not give strong evidence against the ambulance service's claim.

6.23 We need to find

$$\begin{aligned} P(S^2 > 39.74) &= P(\frac{(n-1)S^2}{\sigma^2} > \frac{(n-1)39.74}{\sigma^2}) \\ &= P(\frac{(n-1)S^2}{\sigma^2} > \frac{(14)39.74}{21.32}) = P(\frac{14S^2}{\sigma^2} > 26.12). \end{aligned}$$

Since the data are from a normal population, the statistic $(n-1)S^2/\sigma^2$ has a χ^2 distribution with $n-1$ degrees of freedom. From Table 5, with $\nu = 14$ degrees of freedom, we see that

$$P(\frac{14S^2}{\sigma^2} > 26.12) = .025.$$

Thus, the probability that the claim will be rejected even though $\sigma^2 = 21.3$ is .025 or 2.5 percent.

6.25 We need to find

$$1 - P(\frac{1}{7} \le \frac{S_1^2}{S_2^2} \le 7).$$

Since the samples are independent and from normal populations, the statistic S_1^2/S_2^2 has an F distribution with $n_1 - 1 = n_2 - 1 = 7$ degrees of freedom for both the numerator and the denominator. From Table 6(b), we see that

$$P(\frac{S_1^2}{S_2^2} > 6.99) = .01.$$

Using the relation

$$F_{1-\alpha}(\nu_1, \nu_2) = \frac{1}{F_\alpha(\nu_2, \nu_1)},$$

we know that

$$P(\frac{S_1^2}{S_2^2} < \frac{1}{6.99}) = .01.$$

Thus, approximately

$$P(\frac{S_1^2}{S_2^2} < \frac{1}{7} \; or \; \frac{S_1^2}{S_2^2} > 7)$$

$$= P(\frac{S_1^2}{S_2^2} < \frac{1}{7}) + P(\frac{S_1^2}{S_2^2} > 7) = .01 + .01 + = .02.$$

6.27 We need to find

$$P(S^2 > 180) = P(\frac{(n-1)S^2}{\sigma^2} > \frac{4 \cdot 180}{144}) = P(\frac{(n-1)S^2}{\sigma^2} > 5).$$

Since the sample is from a normal population, the statistic $((n-1)S^2)/(\sigma^2)$ has a χ^2 distribution with $n-1=4$ degrees of freedom. The density of the χ^2 distribution with 4 degrees of freedom is

$$f(x) = \frac{1}{4}xe^{-x/2}, \qquad x > 0.$$

Thus,

$$P(S^2 > 180) = P(\frac{(n-1)S^2}{\sigma^2} > 5) \; = \; \int_5^\infty \frac{1}{4}xe^{-x/2}\,dx.$$

Integrating by parts with $u = x$ and $dv = e^{-x/2}$ gives

$$P(S^2 > 180) = -e^{-x/2}(\frac{1}{2}x + 1) \mid_5^\infty = e^{-5/2}(\frac{5}{2} + 1) = e^{-5/2}\frac{7}{2} = .2873.$$

6.29 The probability that the ratio of the larger to the smaller sample variance exceeds 3 is

$$1 - P(\frac{1}{3} \leq \frac{S_1^2}{S_2^2} \leq 3) = 1 - \int_{1/3}^{3} \frac{6x}{(1+x)^4}\, dx.$$

Let $u = x + 1$ or $x = u - 1$. Then

$$\begin{aligned} P(\frac{1}{3} \leq \frac{S_1^2}{S_2^2} \leq 3) &= \int_{4/3}^{4} \frac{6(u-1)}{u^4}\, du = \frac{-3}{u^2} \Big|_{4/3}^{4} + \frac{2}{u^3} \Big|_{4/3}^{4} \\ &= \frac{24}{16} - \frac{52}{64} = .6875. \end{aligned}$$

Thus,

$$1 - P(\frac{1}{3} \leq \frac{S_1^2}{S_2^2} \leq 3) = 1 - .6875 = .3125.$$

6.31 (a) The number of samples (given that order does not matter) is

$$\binom{8}{2} = \frac{8 \cdot 7}{2 \cdot 1} = 28.$$

(b) The number of samples (given that order does not matter) is

$$\binom{20}{2} = \frac{20 \cdot 19}{2 \cdot 1} = 190.$$

6.33 The finite population correction factor is $(N - n)/(N - 1)$. Thus

(a)

$$\frac{N-n}{N-1} = \frac{8-2}{8-1} = .857.$$

(b)

$$\frac{N-n}{N-1} = \frac{20-2}{20-1} = .947.$$

6.35 Let $\bar{X}$ be the mean number of pieces of mail in a random sample of size 25.

(a) By Chebyshev's theorem, set

$$P(\bar{X} < 40 \text{ or } \bar{X} > 48) = P(|\bar{X} - 44| > 4) \leq \frac{1}{k^2}$$

where

$$4 = k\sigma_{\bar{X}} = k\frac{8}{\sqrt{25}} \quad \text{or} \quad \frac{1}{k^2} = .16$$

Thus,

$$P(\bar{X} < 40 \text{ or } \bar{X} > 48) \leq .16$$

(b) By the central limit theorem,

$$\begin{aligned} P(\bar{X} < 40 \text{ or } \bar{X} > 48) &= 1 - P(40 \leq \bar{X} \leq 48) \\ &= 1 - P(\frac{40-44}{8/\sqrt{25}} \leq \frac{\bar{X}-\mu}{\sigma/\sqrt{n}} \leq \frac{48-44}{8/\sqrt{25}}) \\ &\approx 1 - P(-2.5 \leq Z \leq 2.5) = 2P(Z > 2.5) \\ &= 2(1 - .9938) = .0124 \end{aligned}$$

6.37 By the central limit theorem,

$$\begin{aligned} P(|\bar{X} - \mu| > .06) &= P(\frac{|\bar{X}-\mu|}{\sigma/\sqrt{n}} > \frac{.06}{.16/\sqrt{36}}) \\ &\approx P(|Z| > 2.25) = 2(1 - .9878) = .0244 \end{aligned}$$

6.39 By Theorem 6.5, $F = S_2^2/S_1^2$ has F distribution with 7 and 25 degrees of freedom. Hence , from Table 6(a), we have

$$P(\frac{S_2^2}{S_1^2} > 2.4) = P(F > 2.4) = .05$$

6.41 Under this sampling scheme, the observations will not satisfy the independence assumption required for a random sample. The longest lines are likely to occur together when the same cars must wait several light changes during the rush hour.

6.43 (a) No, we would expect the weight of the bags to vary because the weights of the individual apples vary. With X_i the weight of the $i-th$ apple placed in the bag, $\sum_{i=1}^{30} X_i$ is the total weight of the bag. Then,

$$P(\sum_{i=1}^{30} X_i \leq y) = P(\overline{X} \leq y/30)$$

so the probability concerning total weight can be expressed as a probability concerning $\overline{X}$.

(b) Further,

$$P(\overline{X} \leq y/30) = P(\frac{\bar{X}-\mu}{\sigma/\sqrt{n}} \leq \frac{y/30-\mu}{\sigma/\sqrt{n}})$$

By the central limit theorem $\frac{\bar{X}-\mu}{\sigma/\sqrt{n}}$ is nearly standard normal when $n = 30$. Thus, for any specified value y, we can approximate the probability $P(\sum_{i=1}^{30} X_i \leq y)$ by the standard normal probability $P(Z \leq \frac{y/30-\mu}{\sigma/\sqrt{n}})$

(c) With $\mu = .2$ and $\sigma = .03$ pound

$$P(\sum_{i=1}^{30} X_i \geq 6.2) = P(Z \geq \frac{6.2/30 - .2}{.03/\sqrt{30}}) = 1 - F(1.217) = .1118$$

Chapter 7

INFERENCE CONCERNING MEANS

7.1 (a) There are $\binom{5}{3} = 10$ possible samples of size 3.

Sample	Mean	Probability	Sample	Mean	Probability
3,6,9	6	0.10	3,15,27	15	0.10
3,6,15	8	0.10	6,9,15	10	0.10
3,6,27	12	0.10	6,9,27	14	0.10
3,9,15	9	0.10	6,15,27	16	0.10
3,9,27	13	0.10	9,15,27	17	0.10

(b) We verify that the expected value of $\bar{X}$, for a sample of size 3, is

$$E(\bar{X}) = \frac{1}{10}(6+8+12+9+13+15+10+14+16+17) = \frac{120}{10} = 12.$$

which is equal to the mean of the population. That is, it equals

$$\frac{1}{5}(3+6+9+15+27) = \frac{60}{5} = 12$$

7.3 (a) The 12 medians and the 12 means are

Result	Median	Mean	Result	Median	Mean
2,4,6	4	4	3,1,4	3	2.67
5,3,5	5	4.33	5,5,2	5	4
4,5,3	4	4	3,3,4	3	3.33
5,2,3	3	3.33	1,6,2	2	3
6,1,5	5	4	3,3,3	3	3
2,3,1	2	2	4,5,3	4	4

(b) Stem-and-leaf display of the medians. Note the leaf unit = 0.10

2*	0 0
3*	0 0 0 0
4*	0 0 0
5*	0 0 0

Stem-and-leaf display of the means. Note the leaf unit = 0.10

2**	00 67
3**	00 00 33 33
4**	00 00 00 00 00 33
5**	

(c) To suggest why the claim is true, we compare the dot diagram of medians and the dot diagram of means.

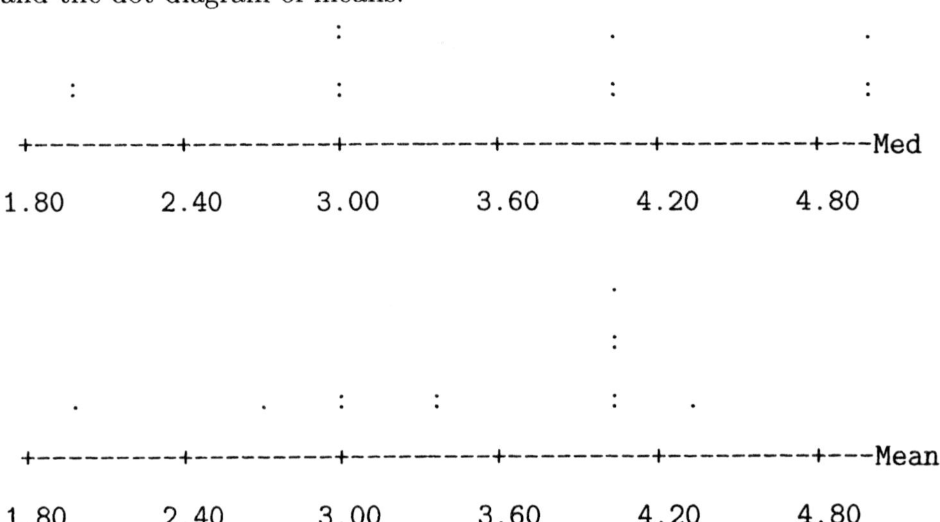

It is clear that, for these samples, the means have less variability then the medians.

7.5 With reference to the Exercise 7.4, the 95 percent confidence interval for the true inter-request time is given by

$$\bar{x} \pm E = 11,795 \pm 3895.57$$

or from 7,899.43 to 15,690.6.

7.7 We need to find $z_{\alpha/2}$ such that

$$.5 = z_{\alpha/2} \cdot \sigma/\sqrt{n}$$

Thus,

$$z_{\alpha/2} = \frac{(.5) \cdot \sqrt{40}}{2.06} = 1.54$$

From Table 3, we find that F(1.54) = .9382. Thus, $\alpha = 2(1 - .9382) = .1236$. Thus, we have about 87.6 percent confidence that the error is less than 30 seconds.

7.9 We need to find $z_{\alpha/2}$ such that

$$10 = z_{\alpha/2} \cdot (62.35)/\sqrt{80}.$$

Thus,

$$z_{\alpha/2} = \sqrt{80} \cdot 10/62.35 = 1.43.$$

From Table 3, we see that F(1.43) = .9236. Thus, $\alpha = 2(1 - .9236) = .1528$. So, the confidence level is about 84.7 percent.

7.11 Since $P\left(|X - \mu| \leq z_{.005}\sigma/\sqrt{n}\right) = .99$, we need to choose n such that

$$z_{.005}\frac{\sigma}{\sqrt{n}} = .25.$$

In this case, $\sigma = 1.40$, $z_{.005} = 2.575$. Thus,

$$n = \left(\frac{(2.575)\cdot 1.40}{.25}\right)^2 = 207.9 \simeq 208.$$

7.13 As discussed in Chapter 6, if a random sample of size n is taken from a population of size N, having mean μ and variance σ^2, then $\bar{X}$ has mean μ and variance $\sigma^2(N-n)/n(N-1)$. Thus, the formula for E becomes

$$E = z_{\alpha/2}\frac{\sigma}{\sqrt{n}}\cdot\sqrt{\frac{N-n}{N-1}}.$$

(a) In this case, $z_{.025} = 1.96$, $s = 85$, $N = 420$, and $n = 50$. Thus,

$$E = 1.96\cdot\frac{85}{\sqrt{50}}\sqrt{\frac{420-50}{420-1}} = 22.14.$$

Thus, we have 95 percent confidence that the error will be less than 22.14.

(b) In this case, $z_{.005} = 2.575$, $\sigma = 12.2$, $n = 40$, $N = 200$. Thus,

$$E = 2.575\frac{12.2}{\sqrt{40}}\cdot\sqrt{\frac{200-40}{200-1}} = 4.454.$$

Thus, we have 99 percent confidence that the error will be less than 4.454.

7.15 A $100(1-\alpha)$ percent confidence interval for the mean is given by

$$\bar{x} \;-\; z_{\alpha/2}\frac{\sigma}{\sqrt{n}} \;<\; \mu \;<\; \bar{x} \;+\; z_{\alpha/2}\frac{\sigma}{\sqrt{n}}$$

In this case, $472 = 487 - 15$ and $502 = 487+15$, $\sigma = 48$, and n $= 100$. Thus, we need to find $z_{\alpha/2}$ such that

$$z_{\alpha/2}\cdot 48/\sqrt{100} = 15.$$

So, $z_{\alpha/2} = 3.125$. Using Table 3, we see that $\alpha = .0018$. Thus, we have 99.82 percent confidence.

7.17 Since the sample size $n = 29$ can be considered large, we can use the large sample confidence interval with $\overline{x} = 1.4707$, $s = 0.5235$, $n = 29$. Since $z_{.05} = 1.645$, the 90 percent confidence interval is

$$1.4707 - (1.645)\frac{.5235}{\sqrt{29}} < \mu < 1.4707 + (1.645)\frac{.5235}{\sqrt{29}}$$

or, $1.31 < \mu < 1.63$. We are 90% confident that the mean cost to rebuild a traction motor is between 1.31 and 1.63 thousand dollars.

7.19 Since the sample size $n = 40$ is large, we can use the large sample confidence interval. A computer calculation gives $\overline{x} = 5.775$ and $s = 3.059$, $n = 40$. Since $z_{.025} = 1.96$, the 95 percent confidence interval is

$$5.775 - (1.96)\frac{3.059}{\sqrt{40}} < \mu < 5.775 + (1.96)\frac{3.059}{\sqrt{40}}$$

or, $4.827 < \mu < 6.723$. We are 95% confident that the mean number of defects is between 4.827 and 6.723.

7.21 Since the data are a small sample from a normal population, we use the small sample confidence interval for μ with $\bar{x} = .5060$, $s = .0040$, $n = 10$, and $t_{.025}$ with 9 degrees of freedom, which is equal to 2.262. Thus, the 95 percent confidence interval is

$$.5060 - (2.262)\frac{.0040}{\sqrt{10}} < \mu < .5060 + (2.262)\frac{.0040}{\sqrt{10}}$$

or, $.5031 < \mu < .5089$.

7.23 In this case, $t_{.01}$ with 5 degrees of freedom is 3.365. Thus, the 98 percent confidence interval is given by

$$4 - \frac{(3.365)(3.162)}{\sqrt{6}} \quad < \quad \mu \quad < \quad 4 + \frac{(3.365)(3.162)}{\sqrt{6}}$$

or, $-.344 \quad < \quad \mu \quad < \quad 8.344$.

7.25 (a) The Poisson probability distribution is

$$f(x;\lambda) = \frac{\lambda^x e^{-\lambda}}{x!} \quad \text{for } x = 0, 1, 2, ...$$

So, the likelihood of a sample $x_1, ..., x_n$ is equal to

$$\prod_1^n \left(\frac{\lambda^{x_i} e^{-\lambda}}{x_i!} \right) = \frac{\lambda^{\sum_1^n x_i} e^{-n\lambda}}{\prod_1^n x_i!}$$

If $\sum_1^n x_i = 0$ then $e^{-\lambda}$ has a maximum at 0. Otherwise, differentiating with respect to λ, we get

$$\frac{\lambda^{\sum_1^n x_i - 1} e^{-n\lambda}}{\prod_1^n x_i!} \left(\sum_1^n x_i - n\lambda \right)$$

which is positive for λ less than $\bar{x}$, equal to zero for $\lambda = \bar{x}$, and negative for λ greater then $\bar{x}$. Thus, the maximum likelihood estimator is $\hat{\lambda} = \bar{X}$ which also includes the case $\bar{X} = 0$.

(b) The likelihood is

$$\prod_1^n p^{x_i}(1-p)^{1-x_i} = p^{\sum_1^n x_i}(1-p)^{n-\sum_1^n x_i}$$

Note that the maximum likelihood estimator $\hat{p} = 0$ if $\sum_1^n x_i = 0$ and $\hat{p} = 1$ if $\sum_1^n x_i = n$ so $\hat{p} = \bar{X}$ in these cases. Otherwise, differentiating the likelihood with respect to p gives

$$p^{\sum_1^n x_i - 1}(1-p)^{n-\sum_1^n x_i - 1}[\sum_1^n x_i - np]$$

so the maximum likelihood estimator is $\hat{p} = \bar{X}$ hold for all cases.

7.27 The firm commits a Type I error if it erroneously rejects the null hypothesis that the dam is safe. If it erroneously accepts the null hypothesis that the dam is safe, it commits a Type II error.

7.29 We can assume from past experience that the standard deviation of the drying times is 2.4 minutes. The null hypothesis is that the mean $\mu = 20$. We reject the null hypothesis if $\bar{X} > 20.50$ minutes.

(a) the probability of a Type I error is the probability that $\bar{X} > 20.50$ when $\mu = 20$. Using a normal approximation to the distribution of the sample mean, this probability is given by

$$1 - F\left(\frac{20.50 - 20}{2.4/\sqrt{36}}\right) = 1 - F(1.25) = 1 - .8944 = .1056$$

(b) The probability of a Type II error when $\mu = 21$ is the probability that $\bar{X} < 20.50$ when $\mu = 21$. Using a normal approximation to the distribution of the sample mean, this probability is given by

$$F\left(\frac{20.50 - 20}{2.4/\sqrt{36}}\right) = F(-1.25) = .1056.$$

7.31 (a) Proceeding as in Exercise 7.29, the probability of a Type I error is given by

$$1 - F\left(\frac{20.75 - 20}{2.4/\sqrt{50}}\right) = 1 - F(2.21) = .014$$

(b) The probability of a Type II error is

$$F\left(\frac{20.75 - 21}{2.4/\sqrt{50}}\right) = F(-.737) = .2327.$$

7.33 (a) The probability of a Type I error is the probability that $\bar{X} < 78.0$ when

$\mu = 80.0$. Using the normal approximation to the distribution of the sample mean, this probability is given by

$$F\left(\frac{78.0 - 80.0}{8.4/\sqrt{100}}\right) = F(-2.38) = .0087$$

(b) The answer is the same since for a composite null hypothesis α is the maximum α for all possible values of the null hypothesis. The maximum occurs when $\mu = 80.0$.

7.35 Using the same formula as in Exercise 7.34 with $z_{\alpha/2}$ gives

$$n = 16^2(2.575 + .8415)^2/(100 - 92)^2 = 46.7 \simeq 47.$$

7.37 (a) In this case, use the two-sided alternative $\mu \neq \mu_0$ where μ is the true mean daily inventory under the new marketing policy, and μ_0 is the true mean daily inventory under the old policy (Note: $\mu_0 = 1250$).

(b) The burden of proof is on the new policy. Thus, the alternative is $\mu_0 > \mu$.

(c) The burden of proof is on the old policy. Thus, the alternative is $\mu_0 < \mu$.

7.39 1. *Null hypothesis* $H_0 : \mu = 73.2$

Alternative hypothesis $H_1 : \mu > 73.2$

2. *Level of significance:* $\alpha = 0.01$.

3. *Criterion:* Using a normal approximation for the distribution of the sample mean, we reject the null hypothesis when

$$Z = \frac{\bar{X} - \mu_0}{\sigma/\sqrt{n}} > z_\alpha.$$

Since $\alpha = .01$ and $z_{.01} = 2.33$, the null hypothesis must be rejected if $Z > 2.33$.

4. *Calculations:* $\mu_0 = 73.2$, $\bar{x} = 76.7$, $\sigma = 8.6$, and $n = 45$ so

$$Z = \frac{76.7 - 73.2}{8.6/\sqrt{45}} = 2.73$$

5. *Decision:* Because $2.73 > 2.33$, the null hypothesis that $\mu = 73.2$ is rejected. at level .01. The P-value $= P[Z > 2.73] = .0032$ as shown in the figure. The evidence against the null hypothesis, $\mu = 73.2$, is very strong.

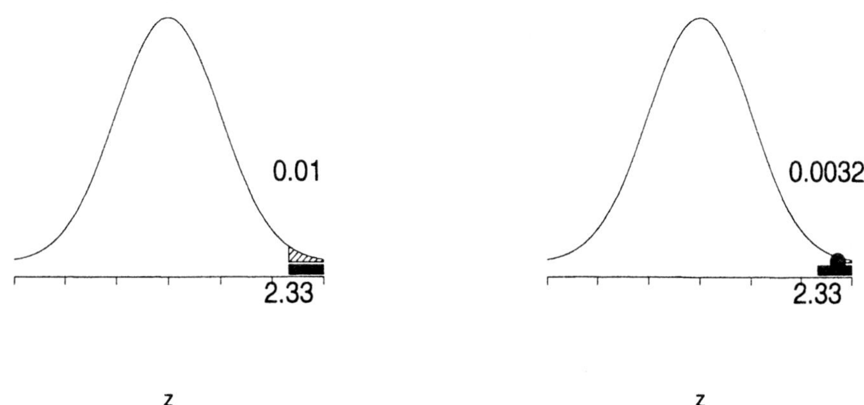

(a) Rejection Region (b) P-value for Problem 7.39

7.41 1. *Null hypothesis* $H_0 : \mu = 1000$

Alternative hypothesis $H_1 : \mu > 1000$

2. *Level of significance:* $\alpha = 0.05$.

3. *Criterion:* Since the sample is large, we will use the normal approximation to the distribution of the mean substituting S for σ. We reject the null hypothesis when

$$Z = \frac{\bar{X} - \mu_0}{S/\sqrt{n}} > z_\alpha.$$

Since $\alpha = .05$ and $z_{.05} = 1.645$, the null hypothesis must be rejected if $Z > 1.645$.

4. *Calculations:* $\mu_0 = 1000$, $\bar{x} = 1038$, $s = 146$, and $n = 64$

$$Z = \frac{1038 - 1000}{146/\sqrt{64}} = 2.08$$

5. *Decision:* Because $2.08 > 1.645$, the null hypothesis that $\mu = 1000$ is rejected. at level .05. The P-value $= P[Z > 2.08] = .0188$ as shown in the figure. The evidence against the null hypothesis, $\mu = 1000$, is quite strong.

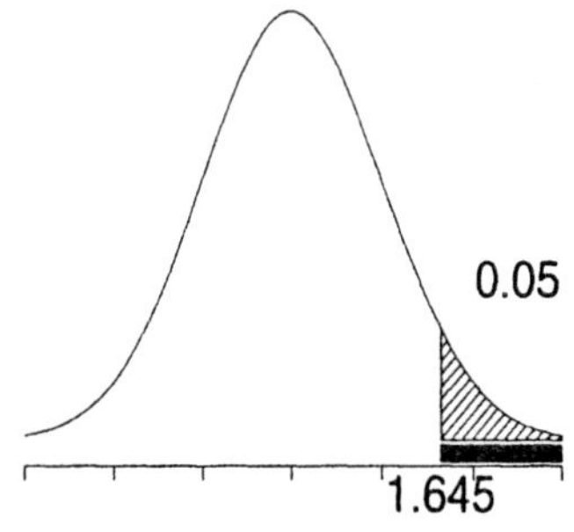

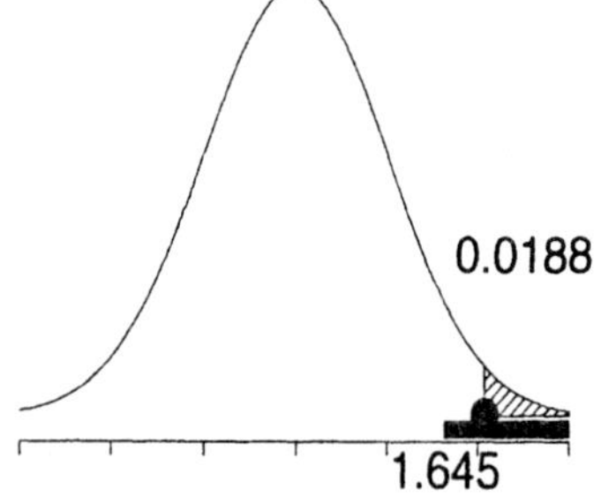

7.43 1. *Null hypothesis* $H_0 : \mu = 32.6$

Alternative hypothesis $H_1 : \mu > 32.6$

2. *Level of significance:* $\alpha = 0.05$.

3. *Criterion:* Since the sample is large, we use the normal approximation to the distribution of the mean. We reject the null hypothesis when

$$Z = \frac{\bar{X} - \mu_0}{S/\sqrt{n}} > z_\alpha.$$

Since $\alpha = .05$ and $z_{.05} = 1.645$, the null hypothesis must be rejected if $Z > 1.645$.

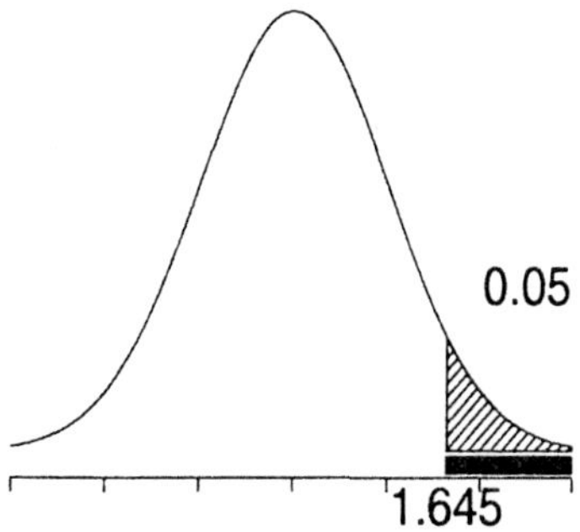

4. *Calculations:* $\mu_0 = 32.6$, $\bar{x} = 33.8$, $s = 6.1$, and $n = 60$ so

$$z = \frac{33.8 - 32.6}{6.1/\sqrt{60}} = 1.52$$

5. *Decision:* Because $1.52 < 1.645$, we cannot reject the null hypothesis at the .05 level of significance.

7.45 1. *Null hypothesis* $H_0 : \mu = 14$

Alternative hypothesis $H_1 : \mu \neq 14$

2. *Level of significance:* $\alpha = 0.01$.

3. *Criterion:* Since the sample is small, we cannot use the normal approximation for $\overline{X}$. If it is reasonable to assume that the data are from a distribution that is nearly normal, we can use the t statistic.

$$t = \frac{\bar{X} - \mu_0}{S/\sqrt{n}}$$

Since the alternative hypothesis is one-sided, the critical region is defined by $t > t_{.01}$ where $t_{.01}$ with 4 degrees of freedom is 3.474.

4. *Calculations:* In this case, $\mu_0 = 14,\ \bar{x} = 14.9,\ s = .42$, and $n = 5$ so

$$t = \frac{14.9 - 14}{.42/\sqrt{5}} = 4.79$$

5. *Decision:* Because $4.79 > 3.474$, we reject the null hypothesis, the mean fat content is 14%, at the .01 level of significance. The P-value is less than .005 .

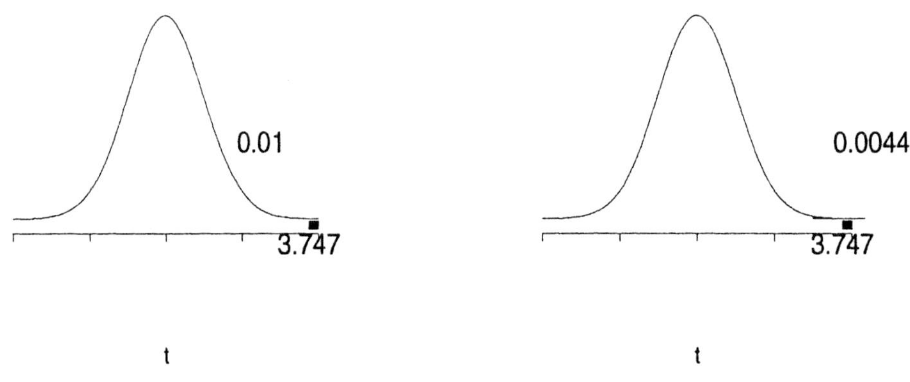

(a) Rejection region (b) P-value for Problem 7.45

7.47 1. *Null hypothesis* $H_0 : \mu = 10.5$

Alternative hypothesis $H_1 : \mu > 10.5$

2. *Level of significance:* $\alpha = 0.01$.

3. *Criterion:* Assuming the population is normal, we can use the t statistic.

$$t = \frac{\bar{X} - \mu_0}{S/\sqrt{n}}$$

Since the alternative hypothesis is one-sided, the critical region is defined by $t > t_{.01}$ where $t_{.01}$ with 7 degrees of freedom is 2.998 .

4. *Calculations:* In this case, $\mu_0 = 10.5$, $\bar{x} = 14$, $s = 3.207$ and $n = 8$ so

$$t = \frac{14 - 10.5}{3.207/\sqrt{8}} = 3.087.$$

5. *Decision:* Because $3.087 > 2.998$, we reject the null hypothesis at the .01 level of significance. A computer program gives the P-value .0088 shown in the figure.

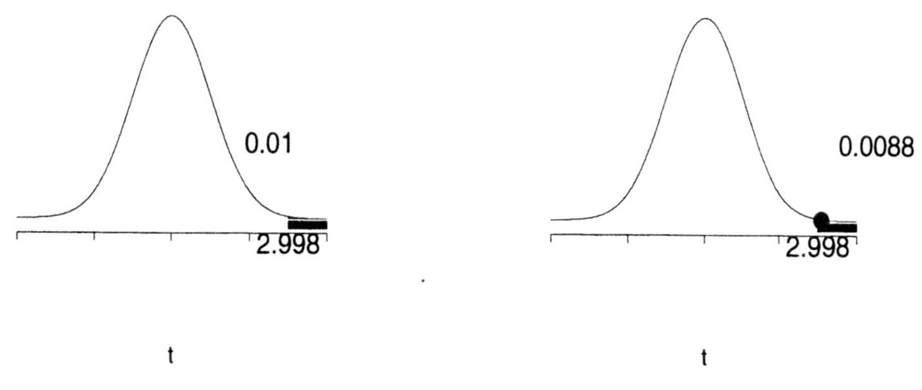

(a) Rejection region (b) P-value for Problem 7.47

7.49 We are testing the null hypothesis $H_0 : \mu = 14.0$ against the alternative $H_1 : \mu \neq 14.0$ at the .05 level of significance. The critical region is defined by $t < -t_{.025}$ or $t > t_{.025}$ where $t_{.025}$ with 4 degrees of freedom is 2.776. With the first value changed, $\bar{x} = 14.7$ and $s = .74162$ so

$$t = \frac{14.7 - 14.0}{.74162/\sqrt{5}} = 2.11.$$

and we cannot reject the null hypothesis. The "paradox" is explained by the standard deviation, which has greatly increased.

7.51 (a) $\mu_0 = 75.2$, $\sigma = 3.6$, so in this case,

$$d = \frac{|76.0 - 75.2|}{3.6} = .222.$$

From Table 8(a) with $n = 15$, we see that the probability of a Type II error is about .80.

(b) $\mu_0 = 75.2$, $\sigma = 3.6$, $\mu = 78.0$, so in this case,

$$d = \frac{|78.0 - 75.2|}{3.6} = .778.$$

From Table 8(a) with $n = 15$, we see that the probability of a Type II error is about .10.

7.53 (a) $\mu_0 = 2.000$, $\mu = 2.020$, $\sigma = .050$, $\alpha = .05$, $n = 30$. Thus,

$$d = \frac{|2.020 - 2.000|}{.050} = .4.$$

Since this is a two-sided test, we use Table 8(c). The probability of a Type II error is about .26.

(b) $\mu = 2.030$. Thus,

$$d = \frac{|2.030 - 2.000|}{.050} = .6.$$

The probability of a Type II error is about .15.

(c) $\mu = 2.040$. Thus,

$$d = \frac{|2.040 - 2.000|}{.050} = .8.$$

The probability of a Type II error is about .05.

7.55 Now the alternative is two-sided, so we use Table 8(d). The probabilities of a Type II errors are about

(a) .92 (b) .79 (c) .56 (d) .30 (e) .12 (f) .05

7.57 Since the alternative is one-sided and $\alpha = .05$, we use Table 8(a) with $n = 8$. The probabilities of a Type II error are about

(a) .59 (b) .32 (c) .12 (d) .03 (e) .00

7.59 $\mu_0 = 100$, $\mu = 92$, $\sigma = 16$, $\alpha = .01$ and $P(\text{Type II error}) = .20$. Since the alternative hypothesis is two-sided, we use Table 8(c) with $d = .5$. We must take a sample size of about 30.

7.61 (a) Since 21.5 is outside the 95 percent confidence interval, the test in Exercise 7.60 is consistent with the confidence interval.

(b) With the sulfur emission data in C1, the MINITAB output is

```
TINTERVAL 90 PERCENT C1

      N    MEAN    STDEV  SE MEAN    90.0 PERCENT C.I.
C1   80  18.896    5.656    0.632  (  17.843,  19.949)
```

(c) With the aluminum alloy data in C2, the MINITAB output is

```
TINTERVAL 95 C2

      N    MEAN    STDEV  SE MEAN    95.0 PERCENT C.I.
C2   58  70.697    1.797    0.236  (  70.224,  71.169)
```

7.63 (a) Since we are interested in the mean and standard deviation of differences, we use Theorem 7.1 to find that the mean difference is $.255 - .249 = .006$ and the standard deviation of the difference is

$$\sqrt{.003^2 + .002^2} = .0036.$$

(b) The probability that the shaft will not fit inside the bearing is the probability that a normal random variable with mean .006 and standard deviation .0036 is less than 0. This is given by

$$F\left(\frac{.000 - .006}{.0036}\right) = F(-1.67) = .0475$$

7.65 1. *Null hypothesis* $H_0 : \mu_1 - \mu_2 = 30$

Alternative hypothesis $H_1 : \mu_1 - \mu_2 > 30$

2. *Level of significance:* $\alpha = 0.01$.

3. *Criterion:* The null hypothesis specifies $\delta = \mu_1 - \mu_0 = 30$. Since the samples are large, we use the large sample statistic where we estimate each population variance by its sample variance

$$Z = \frac{\bar{X}_1 - \bar{X}_2 - \delta}{\sqrt{\frac{S_1^2}{n_1} + \frac{S_2^2}{n_2}}}$$

The alternative is one-sided so we reject the null hypothesis for $Z > z_{.01} = 2.33$

4. *Calculations:* Since $n_1 = 60$, $n_2 = 60$, $\bar{x}_1 = 585.00$, $\bar{x}_2 = 532.20$, $s_1 = 31.20$, and $s_2 = 36.40$

$$z = \frac{585.00 - 532.20 - 30}{\sqrt{31.20^2/60 + 36.40^2/60}} = 3.68$$

5. *Decision:* Because $3.68 > 2.33$, we reject the null hypothesis at the .01 level of significance. Men earn an average of more than thirty dollars per week than the women.

The P-value $P[Z > 3.68]$, about .0002 gives even stronger support for rejecting the null hypothesis.

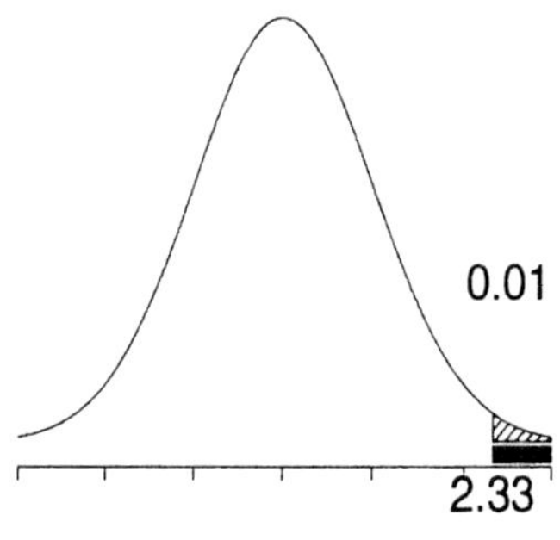

(a) Rejection region

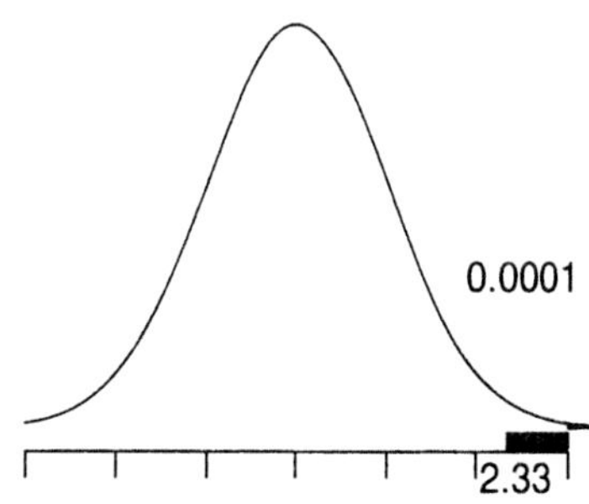

(b) P-value for Problem 7.65

7.67 1. *Null hypothesis* $H_0 : \mu_1 - \mu_2 = 1.5$

Alternative hypothesis $H_1 : \mu_1 - \mu_2 > 1.5$

2. *Level of significance:* $\alpha = 0.05$.

3. *Criterion:* The null hypothesis specifies $\delta = \mu_1 - \mu_0 = 0$. Since the samples are small, but we can assume that the populations are normal with the same variance, we use the two-sample t statistic

$$t = \frac{(\bar{X}_1 - \bar{X}_2) - \delta}{\sqrt{(n_1 - 1)S_1^2 + (n_2 - 1)S_2^2}} \sqrt{\frac{n_1 n_2 (n_1 + n_2 - 2)}{n_1 + n_2}}$$

Since the alternative hypothesis is one-sided, $\delta > 1.5$, we reject the null hypothesis when $t > t_{.05}$ where, from Table 4, with $n_1 + n_2 - 2 = 16$ degrees of freedom, $t_{.05} = 1.746$.

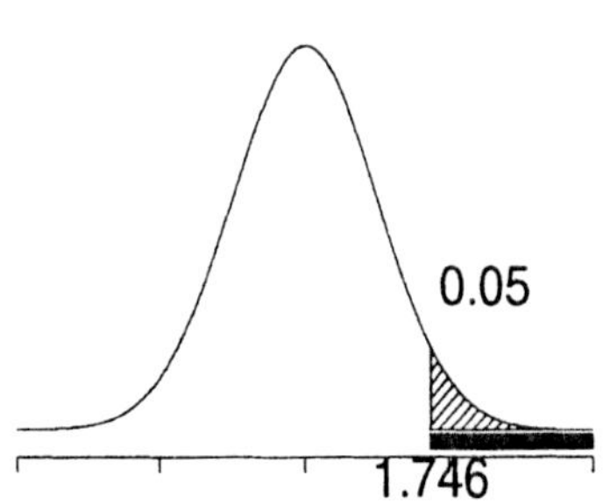

4. *Calculations:* Since $n_1 = 8$, $n_2 = 10$, $\bar{x}_1 = 9.67$, $\bar{x}_2 = 7.43$, $s_1 = 1.81$, and $s_2 = 1.48$

$$t = \frac{(9.67 - 7.43) - 1.5}{\sqrt{7(1.81)^2 + 9(1.48)^2}} \sqrt{\frac{8 \cdot 10 \cdot 16}{18}} = .96,$$

5. *Decision:* We cannot reject the null hypothesis.

7.69 1. Let μ_1 be the mean for California and μ_2 the mean for Oregon. *Null hypothesis* $H_0 : \mu_1 - \mu_2 = 0$

Alternative hypothesis $H_1 : \mu_1 - \mu_2 \neq 0$

2. *Level of significance:* $\alpha = 0.01$.

3. *Criterion:* The null hypothesis specifies $\delta = \mu_1 - \mu_0 = 0$. Since the samples are small, but we can assume that the populations are normal with the same variance, we use the two-sample t statistic

$$t = \frac{(\bar{X}_1 - \bar{X}_2) - \delta}{\sqrt{(n_1 - 1)S_1^2 + (n_2 - 1)S_2^2}} \sqrt{\frac{n_1 n_2 (n_1 + n_2 - 2)}{n_1 + n_2}}$$

Since the alternative hypothesis is two-sided, we reject the null hypothesis when $t < -t_{.005}$ or $t > t_{.005}$ where $t_{.005} = 3.012$ for 13 degrees of freedom.

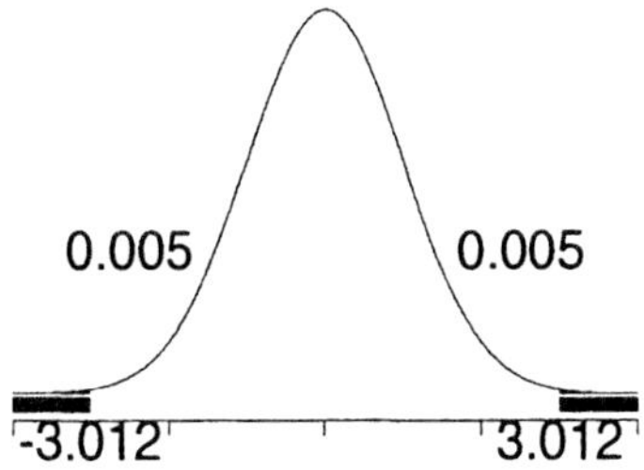

4. *Calculations:* Here $n_1 = 9$ and $n_2 = 6$, and we first calculate $\bar{x}_1 = 58$, $s_1 = 109$, $\bar{x}_2 = 51.833$, $s_2 = 160.97$. Then

$$t = \frac{58 - 51.833}{\sqrt{8(109) + 5(160.97)}}\sqrt{\frac{9 \cdot 6 \cdot 13}{15}} = 1.03.$$

5. *Decision:* We cannot reject the null hypothesis at level of significance $\alpha = .01$.

7.71 1. *Null hypothesis* $H_0 : \mu = 0$

Alternative hypothesis $H_1 : \mu \neq 0$

2. *Level of significance:* $\alpha = 0.05$.

3. *Criterion:* The number of pairs is small so we must assume that each difference has a normal distribution. We use the paired t statistic

$$t = \frac{\bar{D}_1 - \delta}{S_D/\sqrt{n}}$$

Since the alternative hypothesis is two-sided, we reject the null hypothesis if $|t| > t_{.025} = 2.262$.

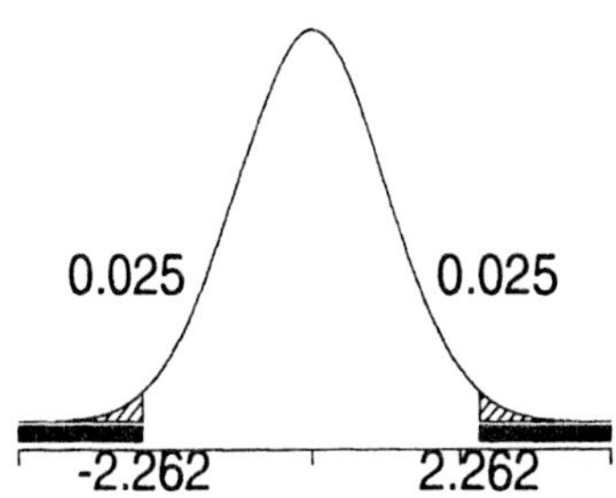

4. *Calculations:* The difference between the weights on Scale I and the weights on Scale II are

$$-.04, -.05, -.02, -.02, .01, -.02, -.05, .01, -.05, .03$$

The mean of this sample is $-.02$ and the variance is $s^2 = .0008$. The null distribution specifies $\delta = 0$ so

$$t = \frac{-.02 - 0}{\sqrt{.0008/9}} = -2.12$$

5. *Decision:* We cannot reject the null hypothesis at level of significance .05.

7.73 Since $\bar{x}_1 - \bar{x}_2 = .053$ and $n_1 = 32$ and $n_2 = 32$, using the large sample confidence interval with $z_{.025} = 1.96$ gives

$$.053 \pm 1.96\sqrt{\frac{.004^2}{32} + \frac{.005^2}{32}} = .053 \pm .002$$

Hence the 95 percent confidence interval is $.051 < \mu < .055$.

7.75 (a) First, order the elevators from 1 to 6. Choose a column from Table 7. Starting somewhere in the column and going down the page. Select 3 different numbers by discarding 0 and numbers larger than 6. These are the numbers for the elevators in which the modified circuit board will be inserted.

(b) For each of the six elevators, toss a coin to decide which board to insert first. If heads, the elevator gets the modified circuit board first. After some time, it is replaced by the original board. If tails, elevator gets the modified circuit board second.

7.77 First, order the cars from 1 to 50. Choose two columns from Table 7. Starting somewhere in the column and going down the page. Select 25 different numbers by discarding 00 and numbers larger than 50. These are the numbered cars in which to install the modified air pollution device.

7.79 The 95% large sample confidence interval for the mean strength of the aluminum alloy 1 is

$$\bar{x} \pm z_{.05/2}\frac{s}{\sqrt{n}} = 70.70 \pm 1.96\frac{1.80}{\sqrt{58}} = 70.70 \pm .46$$

or $70.24 < \mu < 71.16$.

7.81 The 95 percent confidence interval is given by

$$\bar{x} \pm z_{\alpha/2}\frac{\sigma}{\sqrt{n}} = 26.40 \pm 1.48,$$

so the interval is from 24.92 to 27.88.

7.83 Since $t_{.025}$ with 11 degrees of freedom equals 2.201, the 95% confidence interval for the actual mean eccentricity of the can shafts is

$$\bar{x} \pm t_{.025}\frac{s}{\sqrt{n}} = 1.020 \pm 2.201\frac{.044}{\sqrt{12}} = 1.020 \pm .013$$

or $1.007 < \mu < 1.033$.

7.85 First, we use the error bound from the normal distribution to get an initial estimate of the required sample size. Thus, we need to find n_1 such that

$$\frac{(14,380)(1.96)}{\sqrt{n_1}} = 10,000$$

(since $z_{.025} = 1.96$). Thus, $n_1 = 7.94 \simeq 8$. Now we use $t_{.025} = 2.365$ with 7 degrees of freedom to estimate n_2. Thus n_2 is given by

$$\frac{(14,380)(2.365)}{\sqrt{n_2}} = 10,000$$

or, $n_2 = 11.56 \simeq 12$. Next, we use $t_{.025} = 2.201$ with 11 degrees of freedom to estimate n_3. Thus n_3 is given by

$$\frac{(14,380)(2.201)}{\sqrt{n_3}} = 10,000$$

or, $n_3 = 10.017 \simeq 11$. Now, we use $t_{.025}$ with 10 degrees of freedom to find n_4. Proceeding as before, $n_4 = 10.265 \simeq 11$. Since we have converged to a sample of size 11, 11 observations would be required to have 95 percent confidence that the error is less than 10,000.

7.87 The 90% small sample confidence interval for $\mu_1 - \mu_2$ is

$$(.42 - .53) \pm 1.782\sqrt{\frac{6(.0072) + 6(.0044)}{12}}\sqrt{\frac{14}{7 \cdot 7}}$$

$$= -.11 \pm .073 = -.183, \quad -.037$$

or $-.183 < \mu < -.037$.

7.89 The 90% small sample confidence interval for $\mu_1 - \mu_2$ is

$$(.33 - .25) \pm 1.860\sqrt{\frac{2(.0028) + 6(.0032)}{8}}\sqrt{\frac{10}{3 \cdot 7}}$$

$$= .08 \pm .07 = .01,\ .15$$

7.91 (a) The critical region is given by $Z > z_{.03} = 1.8812$ where

$$Z = \frac{\bar{X} - 20}{2.4/\sqrt{50}}$$

Solving for $\bar{X}$ gives the critical region

$$\bar{X} > \frac{(1.8812)(2.4)}{\sqrt{50}} + 20 = 20.6385.$$

(b) The calculation is shown below.

μ	$z = (20.6385 - \mu)/(2.4/\sqrt{50})$	F(z)
19.50	3.354	0.9996
19.75	2.618	0.9956
20.00	1.881	0.9700
20.25	1.145	0.8739
20.50	0.408	0.6584
20.75	−0.329	0.3711
21.00	−1.065	0.1434
21.25	−1.802	0.0358
21.50	−2.538	0.0056
21.75	−3.275	0.0005
22.00	−4.011	0.0000

A plot of the OC curve for the test with sample size 50 is given in Figure 7.1. The OC curve for both tests are shown in Figure 7.2 where the dotted line corresponds to the test for sample size 50.

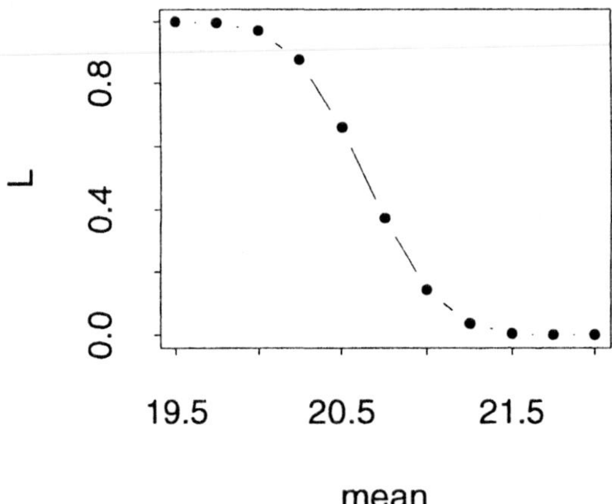

Figure 7.1: OC curve for Exercise 7.91.

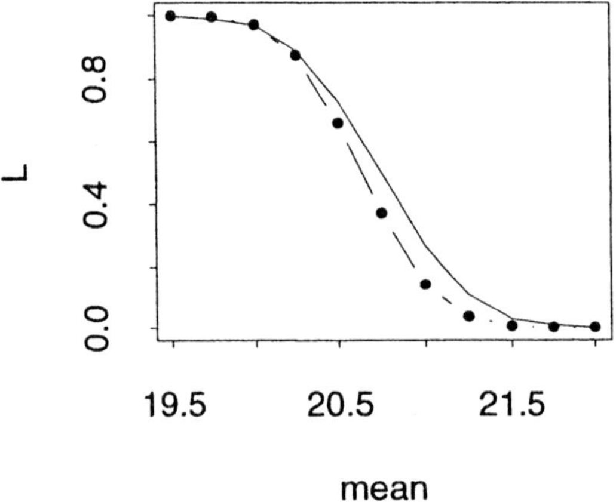

Figure 7.2: OC curve for Exercise 7.91 and Figure 7.5 of text.

7.93 $\mu_0 = 64$, $\mu = 61$, $\sigma = 75$, $\alpha = .05$. Thus, $d = .42$. Since the alternative hypothesis is one-sided, we use Table 8(a) with probability of Type II error .01. The sample size required is about 100.

7.95 Since the alternative hypothesis is one-sided at the $\alpha = .01$ level, we use Table 8b. In this case,

$$d = \frac{|\delta - \delta'|}{\sqrt{\sigma_1^2 + \sigma_2^2}} = \frac{|66.7 - 53.2|}{\sqrt{14.4^2 + 10.8^2}} = .75$$

Thus, the common sample size should be 25.

7.97 (a) Use Table 7 to select 10 cars to install the modified spark plugs. Install the regular plugs to the other cars.

(b) Number the specimens from 1 to 15 and use Table 7 to select 8 numbers between 1 and 20. Use the new oven to bake the specimens having these numbers. Bake the other specimens in the old oven.

7.99 The MINITAB output, for the one sided-test using the data in 7.68, is

```
Two-sample T for Method A vs Method B

N          Mean        StDev    SE Mean
Method A   10        70.00         3.37         1.1
Method B   10        74.00         5.40         1.7

Difference = mu Method A - mu Method B
Estimate for difference:   -4.00
95% upper bound for difference: -0.51
T-Test of difference = 0 (vs <): T-Value = -1.99
P-Value = 0.031   DF = 18
Both use Pooled StDev = 4.50
```

Since the P-value (=0.031) is less than .05, we reject the null hypothesis with 95% of confidence. The conclusion is the same as in Exercise 7.68: method B is more effective.

Chapter 8

INFERENCES CONCERNING VARIANCES

8.1 (a) The sample variance s^2 is given by

$$s^2 = \frac{1}{n-1}\sum(x_i - \bar{x})^2 \;=\; 17.6$$

Thus, the sample standard deviation is $\sqrt{17.6} = 4.195$.

(b) The minimum observation is 21. The maximum observation is 32. thus, the range is 11. Since the sample size is 6, the expected length of the range is 2.534σ. Thus, we estimate σ by $11/2.534 = 4.341$.

8.3 (a) The sample standard deviation is 1.787.

(b) The range of the data is $68.4 - 61.8 = 6.6$. The sample size is 10, so the estimate of the standard deviation is $6.6/3.078 = 2.144$. The relative difference is

$$\frac{2.144 - 1.787}{1.787} \cdot 100 = 20.0 \text{ percent.}$$

8.5 The sample variance is .025. Since the sample size is 5, the $\chi^2_{.01/2}$ with 4 degrees of freedom is 14.860 and $\chi^2_{1-.01/2}$ with 4 degrees of freedom is .207. Thus, if the data are from a normal population, a 99 percent confidence interval for σ^2 is

$$\frac{4(.025)}{14.860} < \sigma^2 < \frac{4(.025)}{.207}$$

or

$$.0067 < \sigma^2 < .483,$$

and the 99 percent confidence interval for the standard deviation is

$$.082 < \sigma < .695$$

8.7 Assuming a normal population, we use the statistic

$$\chi^2 = \frac{(n-1)S^2}{\sigma_0^2}.$$

Since the alternative is $\sigma > 600$, we reject the null hypothesis $\sigma = 600$ if $\chi^2 > \chi^2_{.05}$ with 5 degrees of freedom or if $\chi^2 > 11.076$. In this case, $s = 648$ so the test statistic

$$\chi^2 = \frac{5 \cdot (648)^2}{600^2} = 5.832.$$

Thus, we cannot reject the null hypothesis at the .05 level of significance.

8.9 Since the data are from a normal population, we can use the statistic

$$\chi^2 = \frac{(n-1)S^2}{\sigma_0^2}.$$

The null hypothesis is $\sigma = 15.0$ and the alternative is $\sigma > 15.0$. Since the sample size is 71, we reject the null hypothesis if $\chi^2 > \chi^2_{.05}$ with 70 degrees of freedom Thus, we reject the null hypothesis $\sigma = 15.0$ if $\chi^2 > 90.531$. In this case, $s = 19.3$

minutes and the test statistic

$$\chi^2 = \frac{70 \cdot (19.3)^2}{(15.0)^2} = 115.886$$

so we reject the null hypothesis $\sigma = 15.0$ in favor of the alternative $\sigma > 15.0$, at the .05 level of significance.

8.11 The sample standard deviation is $s = 1.32$ and the sample size is 10. If the data are from a normal population, we can use the statistic

$$\chi^2 = \frac{(n-1)S^2}{\sigma_0^2}.$$

The null hypothesis is $\sigma = 1.20$ and the alternative is $\sigma > 1.20$. Thus, we reject the null hypothesis when $\chi^2 > \chi^2_{.05}$ with 9 degrees of freedom or when $\chi^2 > 16.919$. In this case $s = 1.32$ so

$$\chi^2 = \frac{9 \cdot (1.32)^2}{(1.20)^2} = 10.89$$

Thus, we cannot reject the null hypothesis at the .05 level of significance.

8.13 Since Exercise 7.67 states that the two samples can be assumed to be from normal populations, we can use the statistic

$$F = \frac{S_M^2}{S_m^2}$$

which has an F distribution with $n_M - 1$ and $n_m - 1$ degrees of freedom. The null hypothesis is $\sigma_1^2 = \sigma_2^2$ and the alternative hypothesis is $\sigma_1^2 \neq \sigma_2^2$. The sample sizes are $n_M = 8$ and $n_m = 10$. Thus, we reject the null hypothesis when $F > F_{.02/2}(7,9)$ or when $F > 5.61$. In this case $s_M^2 = (1.81)^2$ and $s_m^2 = (1.48)^2$ so

$$F = \left(\frac{1.81}{1.48}\right)^2 = 1.496.$$

Thus, we cannot reject the null hypothesis at the .02 level of significance.

8.15 The null hypothesis $\sigma_1^2 = \sigma_2^2$ and the alternative hypothesis is $\sigma_1^2 < \sigma_2^2$. Therefore, we use the statistic

$$F = \frac{S_2^2}{S_1^2}$$

which has an F distribution with $n_2 - 1 = 20$ and $n_1 - 1 = 14$ degrees of freedom. The null hypothesis $\sigma_1^2 = \sigma_2^2$ will be rejected favor of the alternative hypothesis $\sigma_1^2 < \sigma_2^2$ if $F > F_{.01}(20, 14) = 3.51$. In this case

$$F = \left(\frac{4.2}{2.7}\right)^2 = 2.42$$

so we cannot reject the null hypothesis at the .01 level of significance. This analysis assumes that the two samples come from normal populations and that the samples are independent.

8.17 (a) The two samples are not from normal populations so we cannot directly use a two-sample t test. Also, the population variances may be unequal.

(b) If we take the logarithm of each observation, then the transformed data are samples from two normal populations. On this scale, we can then apply the F test for the equality of variances. Specifically, we would test the null hypothesis $\sigma_1^2 = \sigma_2^2$ versus the alternative hypothesis $\sigma_1^2 \neq \sigma_2^2$ using the logarithms of the original data.

8.19 For alloy 1 the sample size is 58 and $\chi^2_{.05/2}$ with 57 degrees of freedom is 79.8 and $\chi^2_{1-.05/2}$ with 57 degrees of freedom is 38.0. The sample standard deviation is 1.80. Thus, the 95 percent confidence interval for σ_1 is

$$\sqrt{\frac{57 \cdot (1.80)^2}{79.8}} < \sigma_1 < \sqrt{\frac{57 \cdot (1.80)^2}{38.0}}$$

or

$$1.52 < \sigma_1 < 2.20$$

For alloy 2, the sample size is 27 and the $\chi^2_{1-.05/2}$ with 26 degrees of freedom is 13.844 and $\chi^2_{.05/2}$ with 26 degrees of freedom is 41.923. The sample standard deviation is 2.42. Thus, the 95 percent confidence interval for σ_2 is

$$\sqrt{\frac{26 \cdot (2.42)^2}{41.923}} < \sigma_2 < \sqrt{\frac{26 \cdot (2.42)^2}{13.844}}$$

or

$$1.91 < \sigma_2 < 3.32.$$

8.21 The sample standard deviation is 4.9. The sample size is 25. The null hypothesis is $\sigma^2 = 30.0$ and the alternative is $\sigma^2 < 30.0$. If the data are from a normal population, we can use the statistic

$$\chi^2 = \frac{(n-1)S^2}{\sigma_0^2}.$$

We reject the null hypothesis if $\chi^2 < \chi^2_{1-.05}$ with 24 degrees of freedom or when $\chi^2 < 13.484$. In this case,

$$\chi^2 = \frac{24 \cdot (4.9)^2}{30} = 19.21.$$

Thus, we cannot reject the null hypothesis at the .05 level of significance.

8.23 The sample size is 101 and $\chi^2_{.05/2}$ with 100 degrees of freedom is 129.561 and $\chi^2_{1-.05/2}$ with 100 degrees of freedom is 74.222. The null hypothesis is $\sigma^2 = .18$ and the alternative is $\sigma^2 \neq .18$. The sample variance is .13. If the data are from a normal population, we can use the statistic

$$\chi^2 = \frac{(n-1)S^2}{\sigma_0^2}.$$

We reject the null hypothesis if $\chi^2 > \chi^2_{.05/2} = 129.561$ or when $\chi^2 < \chi^2_{1-.05/2} = 74.222$. In this case,

$$\chi^2 = \frac{(100)(.13)}{.18} = 72.22.$$

Thus, we reject the null hypothesis $\sigma^2 = .18$, in favor of the alternative $\sigma^2 \neq .18$, at the .05 level of significance. The inspector is not making satisfactory measurements.

8.25 The variance of the first sample is 7.499 and the sample size is 10. The variance of the second sample is 2.681 and the sample size is 8. The null hypothesis $\sigma_1^2 = \sigma_2^2$ and the alternative hypothesis is $\sigma_1^2 \neq \sigma_2^2$. Assuming normal populations we use the statistic

$$F = \frac{S_M^2}{S_m^2}$$

which has an F distribution with $n_M - 1 = 9$ and $n_m - 1 = 7$ degrees of freedom. The null hypothesis $\sigma_1 = \sigma_2$ will be rejected in favor of the alternative hypothesis $\sigma_1 \neq \sigma_2$ if $F > F_{.01}(9, 7) = 6.72$. In this case

$$F = \frac{7.499}{2.681} = 2.797$$

so we cannot reject the null hypothesis at the .02 level of significance.

8.27 (a) No. The distribution which produced these observations is clearly not normal. The dot diagram exhibits a long right hand tail.

(b) The natural logarithms look much more normal. They have

$$\sum_{i=1}^{1} 6 \ln(x_i) = 27.2752 \qquad \sum_{i=1}^{1} 6 \ln^2(x_i) = 67.1147$$

so $s = \sqrt{\dfrac{67.1147 - 27.2752^2/16}{15}} = 1.172$. From the chi-square table, with 15 degrees of freedom, $\chi^2_{0.025} = 27.448$ and and $\chi^2_{0.975} = 6.262$ so the 95 %

confidence interval for σ is

$$\sqrt{\frac{15(1.172)^2}{27.448}} < \sigma < \sqrt{\frac{15(1.172)^2}{6.262}}$$

or

$$.87 < \sigma < 1.81$$

Note however that this result cannot be translated to a statement about the variance on the original scale.

Chapter 9

INFERENCES CONCERNING PROPORTIONS

9.1 (a) The sample proportion is .42. Using Table 9(a) for sample size 200 gives the 95% confidence interval for p,

$$.35 < p < .49.$$

(b) The large sample 95% confidence interval for p obtained by substituting $x/n = .42$ and $z_{\alpha/2} = 1.96$ is

$$.42 - 1.96\sqrt{\frac{(.42)(.58)}{200}} < p < .42 + 1.96\sqrt{\frac{(.42)(.58)}{200}}$$

or

$$.352 < p < .488.$$

9.3 (a) The sample proportion is $231/400 = .578$. Using Table 9(a) for sample size

400 gives the 99% confidence interval

$$.52 < p < .64.$$

(b) Using the large sample formula with $x/n = .578$ and $z_{\alpha/2} = 2.575$ gives the 99% confidence interval

$$.578 - 2.575\sqrt{\frac{(.578)(.422)}{400}} < p < .578 + 2.575\sqrt{\frac{(.578)(.422)}{400}}$$

or

$$.514 < p < .642.$$

9.5 (a) The sample proportion is .38. Using Table 9(a) and interpolating between sample sizes 200 and 400 gives

$$.315 + (.33 - .315)/4 < p < .45 - (.45 - .43)/4$$

or

$$.319 < p < .445.$$

(b) Using the large sample formula for the maximum error with $x/n = .38$ and $z_{\alpha/2}$ =1.96 gives

$$E = 1.96\sqrt{\frac{(.38)(.62)}{250}} = .06.$$

Thus, the error is bounded by .06 with 95% confidence.

9.7 The sample proportion is $69/120 = .575$. Using the large sample formula for the maximum error with $z_{\alpha/2}$ =1.96 gives

$$E = 1.96\sqrt{\frac{(.575)(.425)}{120}} = .0885.$$

Thus, the error is less than .0885 with 95% confidence.

9.9 The sample proportion is $204/300 = .68$. Using the large sample confidence interval with $z_{\alpha/2}$ =2.33 gives

$$.68 - 2.33\sqrt{\frac{(.68)(.32)}{300}} < p < .68 + 2.33\sqrt{\frac{(.68)(.32)}{300}}$$

or $.617 < p < .743$ as the 98% confidence interval.

9.11 We use the formula for p 'known' with $p = .75$, $z_{\alpha/2} = 1.96$, and $E = .06$. Thus,

$$n = (.75)(.25)\left(\frac{1.96}{.06}\right)^2 = 200.08.$$

The required sample size is 201.

9.13 Using the formula for 'known' p with $p = .4$, $z_{\alpha/2} = 2.575$, and $E = .035$ gives

$$n = (.4)(.6)\left(\frac{2.575}{.035}\right)^2 = 1299.06.$$

The required sample size is 1300.

9.15 Using the formula in the preceding exercise with $x = 231$, $n = 400$ and $z_{\alpha/2} = 2.575$,

$$p = \frac{231 + 2.575^2/2 \pm 2.575\sqrt{231(400-231)/400 + 2.575^2/4}}{400 + 2.575^2},$$

or

$$p = .5762 \pm .063.$$

Thus, the 99% confidence interval is

$$.513 < p < .639.$$

9.17 The sample proportion is 533/4063 = .131. Using the large sample confidence interval with $z_{\alpha/2}$ =1.96 gives

$$.131 - 1.96\sqrt{\frac{(.131)(.869)}{4063}} < p < .131 + 1.96\sqrt{\frac{(.131)(.869)}{4063}}$$

or $.12 < p < .14$ as the 95% confidence interval.

9.19

1. *Null hypothesis* $H_0 : p = .3$

   Alternative hypothesis $H_1 : p > .3$

2. *Level of significance:* $\alpha = 0.05$.

3. *Criterion:* Using a normal approximation for the binomial distribution, we reject the null hypothesis when

$$Z = \frac{X - np_0}{\sqrt{np_0(1 - p_0)}} > z_{.05}.$$

   Since $\alpha = .05$ and $z_{.05} = 1.645$, the null hypothesis must be rejected if $Z > 1.645$.

4. *Calculations:* $p_0 = .3$, $X = 47$, and $n = 120$ so

$$Z = \frac{47 - 120(.30)}{\sqrt{120(.30)(.70)}} = 2.19.$$

5. *Decision:* Since the observed value $2.19 > z_{.05} = 1.645$, we reject the null hypothesis at the 5% level of significance. The evidence against the null hypotheses is quite strong since the P-value is .0143.

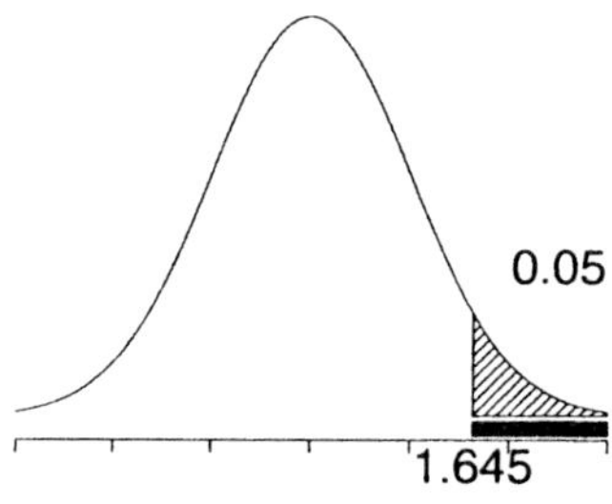

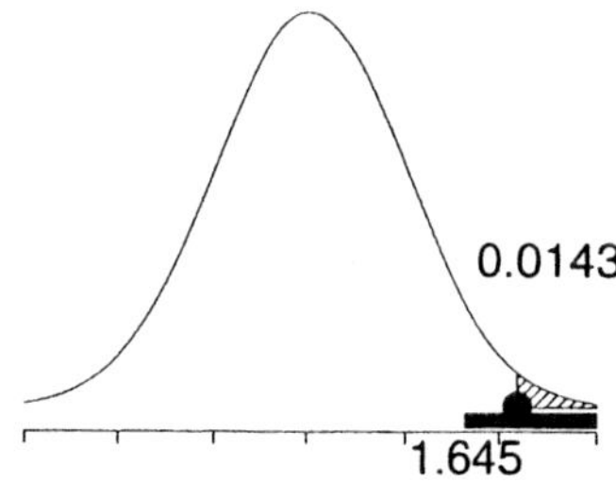

(a) Rejection Region (b) P-value for Problem 9.19

9.21 1. *Null hypothesis* $H_0 : p = .4$

Alternative hypothesis $H_1 : p < .4$

2. *Level of significance:* $\alpha = 0.01$.

3. *Criterion:* Using a normal approximation for the binomial distribution, we reject the null hypothesis when .

$$Z = \frac{X - np_0}{\sqrt{np_0(1 - p_0)}} < -z_{.01}.$$

Since $\alpha = .01$ and $z_{.01} = 2.33$, the null hypothesis must be rejected if $Z < -2.33$.

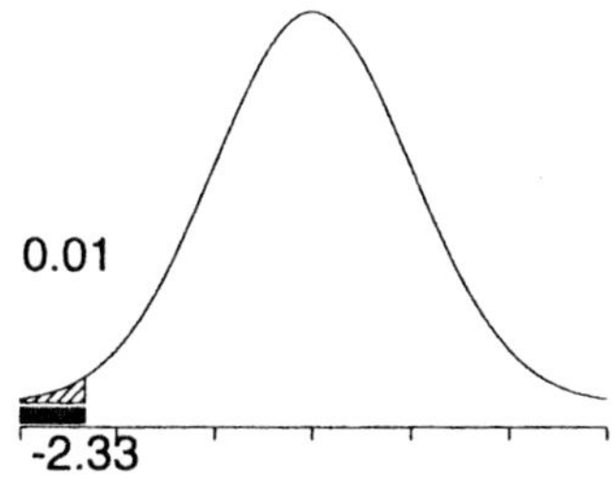

4. *Calculations:* $p_0 = .4$, $X = 49$, and $n = 150$ so

$$Z = \frac{49 - 150(.40)}{\sqrt{150(.40)(.60)}} = -1.83.$$

5. *Decision:* Since the observed value $-1.83 > -z_{.01} = -2.33$, we cannot reject the null hypothesis at the 1% level of significance. We cannot reject the service's claim.

9.23 1. *Null hypothesis* $H_0 : p = .06$

Alternative hypothesis $H_1 : p > .06$

2. *Level of significance:* $\alpha = 0.05$.

3. *Criterion:* Using a normal approximation for the binomial distribution, we reject the null hypothesis when

$$Z = \frac{X - np_0}{\sqrt{np_0(1 - p_0)}} > z_{.05}.$$

Since $\alpha = .05$ and $z_{.05} = 1.645$, the null hypothesis must be rejected if $Z > 1.645$.

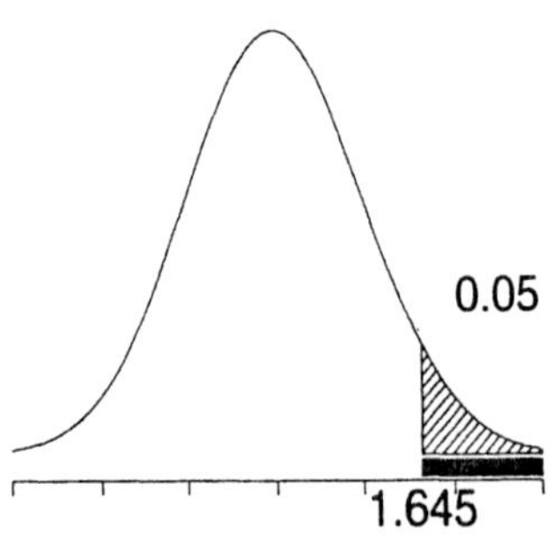

4. *Calculations:* $p_0 = .06$, $X = 17$, and $n = 200$ so

$$Z = \frac{17 - 200(.06)}{\sqrt{200(.06)(.94)}} = 1.489.$$

5. *Decision:* Since the observed value $1.489 < z_{.05} = 1.645$, we cannot reject the null hypothesis at the 5% level.

9.25 The null hypothesis is $p = .50$, the alternative is $p \neq .50$, and the significance level is .05. Thus, we need to find two values x_{α_1} and x_{α_2} where the probability of x_{α_1} or fewer heads plus the probability of x_{α_2} or more heads in 15 flips, when $p = .50$, is less than or equal to .05. Using Table 1 we see that $x_{\alpha_1} = 3$ and $x_{\alpha_2} = 12$ satisfy

$$P[\, X \leq 3 \text{ or } X \geq 12 \mid n = 15, p = .50\,] = .0176 + (1 - .9824) = .0352.$$

The actual level of significance is then .0352.

9.27 We use the χ^2 statistic with 4 degrees of freedom to test the null hypothesis that $p_1 = p_2 = p_3 = p_4 = p_5$ against the alternative that at least one of the probabilities is not equal. Thus, we reject the null hypothesis at the 1% level when $\chi^2 > \chi^2_{.01} = 13.277$. In Table 9.1, the expected frequency and the contribution to the χ^2 statistic of each cell are given in the parentheses and brackets respectively.

The χ^2 statistic is

$$\begin{aligned} \chi^2 &= \frac{.67^2}{14.67} + \frac{3.67^2}{29.33} + \frac{5.40^2}{26.40} + \frac{.60^2}{17.60} + \frac{3.00^2}{22.00} + \\ &\quad \frac{.67^2}{85.33} + \frac{3.67^2}{170.67} + \frac{5.40^2}{153.60} + \frac{.60^2}{102.40} + \frac{3.00^2}{128.00} = 2.37. \end{aligned}$$

Thus, we cannot reject the null hypothesis.

Table 9.1. Exercise 9.27.

	I	II	III	IV	V	Total
Fail	14 (14.67) [.03]	33 (29.33) [.46]	21 (26.40) [1.10]	17 (17.60) [.02]	25 (22.00) [.41]	110
Pass	86 (85.33) [.01]	167 (170.67) [.08]	159 (153.60) [.19]	103 (102.40) [.00]	125 (128.00) [.07]	640
Total	100	200	180	120	150	750

9.29 We use the χ^2 statistic with 2 degrees of freedom to test the null hypothesis that the actual proportions are the same against the alternative that they are not the same. Thus, we reject the null hypothesis at the 1% level when $\chi^2 > \chi^2_{.01} = 9.210$. In Table 9.2, the expected frequency and the contribution to the χ^2 statistic of each cell are given in the parentheses and brackets respectively.

Table 9.2. Exercise 9.29.

	Agency I	Agency II	Agency III	Total
For the pension plan	67 (65.00) [.06]	84 (97.50) [1.87]	109 (97.50) [1.36]	260
Against the pension plan	33 (35.00) [.11]	66 (52.50) [3.47]	41 (52.50) [2.52]	140
Total	100	150	150	400

The χ^2 statistic is

$$\chi^2 = \frac{2.00^2}{65.00} + \frac{13.50^2}{97.50} + \frac{11.50^2}{97.50} + \frac{2.00^2}{35.00} + \frac{13.50^2}{52.50} + \frac{11.50^2}{52.50} = 9.39.$$

Thus, we reject the null hypothesis.

9.31 With reference to part (b) of the preceding exercise,

$$Z^2 = \frac{(13/250 - 7/250)^2}{(.04)(.96)(2/250)} = 1.875.$$

This verifies the fact that $Z^2 = \chi^2$ in this case.

9.33 Let p_1 and p_2 be proportions of reworking units before and after the training respectively. The 99% confidence interval for the true difference of the proportions, $p_1 - p_2$, is

$$x_1/n_1 - x_2/n_2 \pm z_{\alpha/2}\sqrt{\frac{(x_1/n_1)(1-x_1/n_1)}{n_1} + \frac{(x_2/n_2)(1-x_2/n_2)}{n_2}}$$
$$= 26/200 - 12/200 \pm 2.575\sqrt{\frac{(26/200)(1-26/200)}{200} + \frac{(12/200)(1-12/200)}{200}}$$
$$= .07 \pm .075$$

or

$$-.005 < p_1 - p_2 < .145.$$

9.35 The 99% confidence interval for the true difference of the proportions is

$$x_1/n_1 - x_2/n_2 \pm z_{\alpha/2}\sqrt{\frac{(x_1/n_1)(1-x_1/n_1)}{n_1} + \frac{(x_2/n_2)(1-x_2/n_2)}{n_2}}$$
$$= 205/250 - 137/250$$
$$\pm 2.575\sqrt{\frac{(205/250)(1-205/250)}{250} + \frac{(137/250)(1-137/250)}{250}}$$

$$= .272 \pm .102$$

or

$$.170 < p_1 - p_2 < .374.$$

9.37 Notice that $n = \sum_j n_j$ and

$$e_{1j} = n_j \frac{x}{n}, \; e_{2j} = n_j \frac{n-x}{n}.$$

The sum of the expected frequencies of the first row is

$$\sum_j e_{1j} = \sum_j n_j \frac{x}{n} = \frac{x}{n} \sum_j n_j = x.$$

Similarly, the sum of the expected frequencies of the second row is

$$\sum_j e_{2j} = \sum_j n_j \frac{n-x}{n} = \frac{n-x}{n} \sum_j n_j = n - x.$$

Also, the sum of the expected frequencies of the jth column is

$$e_{1j} + e_{2j} = n_j \frac{x}{n} + n_j \frac{n-x}{n} = n_j.$$

9.39 The null hypothesis is that there has been no change of opinion, and the alternative is there has. We use the χ^2 statistic with 2 degrees of freedom to test at the 5% level. Thus, we reject the null hypothesis if $\chi^2 > \chi^2_{.05} = 5.991$. In Table 9.3, the expected frequency and the contribution to the χ^2 statistic of each cell are given in the parentheses and brackets respectively.

Table 9.3. Exercise 9.39.

	2 weeks	4 weeks	Total
Republican Party	79 (85) [.42]	91 (85) [.42]	170
Democratic Party	84 (75) [1.08]	66 (75) [1.08]	150
Undecided	37 (40) [.23]	43 (40) [.23]	80
Total	200	200	400

The χ^2 statistic is

$$\begin{aligned}\chi^2 &= \frac{6^2}{85}+\frac{6^2}{85}+\\ &\quad \frac{9^2}{75}+\frac{9^2}{75}+\\ &\quad \frac{3^2}{40}+\frac{3^2}{40} = 3.457.\end{aligned}$$

Thus, we cannot reject the null hypothesis.

9.41 To test the null hypothesis that there is no dependence between fidelity and selectivity against the alternative that they is dependence at the 1% level, we use the χ^2 statistic with 4 degrees of freedom and reject the null hypothesis when $\chi^2 > \chi^2_{.01} = 13.277$. In Table 9.4, the expected frequency and the contribution to the χ^2 statistic of each cell are given in the parentheses and brackets respectively.

Table 9.4. Exercise 9.41.

Fidelity

Selectivity		Low	Average	High	Total
	Low	6 (13.68) [4.31]	12 (23.16) [5.38]	32 (13.16) [26.98]	50
	Average	33 (30.65) [.18]	61 (51.87) [1.61]	18 (29.47) [4.47]	112
	High	13 (7.66) [3.72]	15 (12.97) [.32]	0 (7.37) [7.37]	28
	Total	52	88	50	190

The χ^2 statistic is

$$\begin{aligned}\chi^2 &= \frac{7.68^2}{13.68} + \frac{11.16^2}{23.16} + \frac{18.84^2}{13.16} + \\ &\frac{2.35^2}{30.65} + \frac{9.13^2}{51.87} + \frac{11.47^2}{29.47} + \\ &\frac{5.34^2}{7.66} + \frac{2.03^2}{12.97} + \frac{7.37^2}{7.37} \\ &= 54.328.\end{aligned}$$

Thus, we reject the null hypothesis and conclude that there is dependence between fidelity and selectivity. The major contributions to χ^2 come from the High Fidelity category. The Low Selectivity and High Fidelity count is very high.

9.43 The mean of the observed distribution is

$$\bar{x} = \frac{0 \cdot 101 + 1 \cdot 79 + 3 \cdot 1}{200} = .6.$$

Thus $.6/4 = .15$ or 15% of the tractors require adjustment. The binomial probabilities and expected frequencies are:

Table 9.5. Exercise 9.43.

No. needing adjustments	Binomial prob. p=.15, n=4	Expected numbers	
0	.5220	104.4	
1	.3685	73.7	
2	.0975	19.5	
3	.0115	2.3	21.9
4	.0005	.1	

The null hypothesis is that the data are from a binomial distribution with $p = .15$ and the alternative is that the data are not from a binomial distribution with $p = .15$. The χ^2 statistic now has 1 degree of freedom, so we reject the null hypothesis at the 1% level when $\chi^2 > \chi^2_{.01} = 6.635$. Now,

$$\chi^2 = \frac{(101 - 104.4)^2}{104.4} + \frac{(79 - 73.7)^2}{73.7} + \frac{(20 - 21.9)^2}{21.9} = .657.$$

Thus, we cannot reject the null hypothesis.

9.45 The mean of the Poisson distribution is 2.0. Using Table 2, the data and expected frequencies are:

Table 9.6. Exercise 9.45.

Number	Frequency		Poisson prob. λ=2.0	Expected frequency	
0	52		.135	67.5	
1	151		.271	135.5	
2	130		.271	135.5	
3	102		.180	90.0	
4	45		.090	45.0	
5	12		.036	18.0	
6	5	} 8	.012	6.0	} 8.5
7	1		.004	2.0	
8	2		.001	.5	

We use the χ^2 statistic with $7-2=5$ degrees of freedom to test the null hypothesis that the arrival distribution is Poisson. Thus, we reject the null hypothesis at the 5% level when $\chi^2 > \chi^2_{.05} = 11.070$. Now,

$$\begin{aligned}\chi^2 &= \frac{(52-67.5)^2}{67.5} + \frac{(151-135.5)^2}{135.5} + \frac{(130-135.5)^2}{135.5} + \frac{(102-90.0)^2}{90.0} + \\ &\frac{(45-45.0)^2}{45.0} + \frac{(12-18.0)^2}{18.0} + \frac{(8-8.5)^2}{8.5} = 9.185.\end{aligned}$$

Thus, we cannot reject the null hypothesis.

9.47 (a) and (b) The calculation of the probability and expected frequency of each class is summarized in the tables:

Table 9.7(a). Exercise 9.47.

Class boundary	Standardized value $=z$	$F(z)$
8.95	-1.78	.0375
12.95	-1.06	.1446
16.95	-0.34	.3669
20.95	0.38	.6480
24.95	1.10	.8643
28.95	1.82	.9656

Table 9.7(b). Exercise 9.47.

Class boundaries	Probability	Expected frequency		Frequency	
$<$ 8.95	.0375	3.0	11.6	3	13
8.95 – 12.95	.1071	8.6		10	
12.95 – 16.95	.2223	17.8		14	
16.95 – 20.95	.2811	22.5		25	
20.95 – 24.95	.2163	17.3		17	
24.95 – 28.95	.1013	8.1	10.9	9	11
$>$ 28.95	.0344	2.8		2	

(c) The χ^2 statistic has $k-3$ degrees of freedom because three things were calculated from the data, namely, the number of the observations, the mean and the standard deviation. In this case, $k=5$, so we reject the null hypothesis at the 5% level when $\chi^2 > \chi^2_{.05} = 5.991$. Now,

$$\chi^2 = \frac{(13-11.6)^2}{11.6} + \frac{(14-17.8)^2}{17.8} + \frac{(25-22.5)^2}{22.5}$$

$$+\frac{(17-17.3)^2}{17.3}+\frac{(11-10.9)^2}{10.9} \quad = \quad 1.264.$$

Thus, we cannot reject the null hypothesis.

9.49 The MINITAB output is:

```
Expected counts are printed below observed counts

      METHOD 1 METHOD 2 METHOD 3    Total
    1       31       42       22       95
         31.67    31.67    31.67

    2       19        8       28       55
         18.33    18.33    18.33

Total       50       50       50      150

ChiSq =  0.014 +  3.372 +  2.951 +
         0.024 +  5.824 +  5.097 = 17.282
df = 2
```

9.51 (a) The sample proportion is .18. Using Table 9(a) for sample size 100 gives the 95% confidence interval for p

$$.115 < p < .275.$$

(b) Using the large sample confidence interval with $x/n = .18$ and $z_{\alpha/2} = 1.96$ gives the 95% confidence interval

$$.18 - 1.96\sqrt{\frac{(.18)(.82)}{100}} < p < .18 + 1.96\sqrt{\frac{(.18)(.82)}{100}}$$

or

$$.105 < p < .255.$$

9.53 (a) The sample proportion is $24/160 = .15$. Using Table 9(b) and interpolating between sample sizes 100 and 200 gives

$$.075 + (3/5)(.09 - .075) < p < .225 + (2/5)(.27 - .225)$$

or

$$.084 < p < .243.$$

(b) Using the large sample confidence interval with $x/n = .15$ and $z_{\alpha/2} = 2.575$ gives the 99% confidence interval

$$.15 - 2.575\sqrt{\frac{(.15)(.85)}{160}} < p < .15 + 2.575\sqrt{\frac{(.15)(.85)}{160}}$$

or

$$.077 < p < .223.$$

9.55 (a) The sample proportion is .13. Using Table 9(a) for sample size 100 gives the 95% confidence interval for p

$$.075 < p < .217.$$

(b) Using the large sample confidence interval with $x/n = .13$ and $z_{\alpha/2} = 1.96$ gives the 95% confidence interval

$$.13 - 1.96\sqrt{\frac{(.13)(.87)}{100}} < p < .13 + 1.96\sqrt{\frac{(.13)(.87)}{100}}$$

or

$$.064 < p < .196.$$

9.57 The large sample 95% confidence bound for p uses the upper limit of the 90 %

confidence interval. Substituting $x/n = 20/4000 = .005$ and $z_{\alpha/2} = 1.645$ gives

$$p < .005 + 1.645\sqrt{\frac{(.005)(.995)}{4000}}$$

or $\qquad p < .0068$

9.59 1. *Null hypothesis* $H_0 : p_1 = p_2$

Alternative hypothesis $H_1 : p_1 > p_2$

2. *Level of significance:* $\alpha = 0.05$.

3. *Criterion:* We using the large sample statistic and reject the null hypothesis when

$$Z = \frac{X_1/n_1 - X_2/n_2}{\sqrt{\hat{p}(1-\hat{p})(\frac{1}{n_1}+\frac{1}{n_2})}} \quad \text{with } \hat{p} = \frac{X_1 + X_2}{n_1 + n_2}$$

is greater than $z_{.05} = 1.645$.

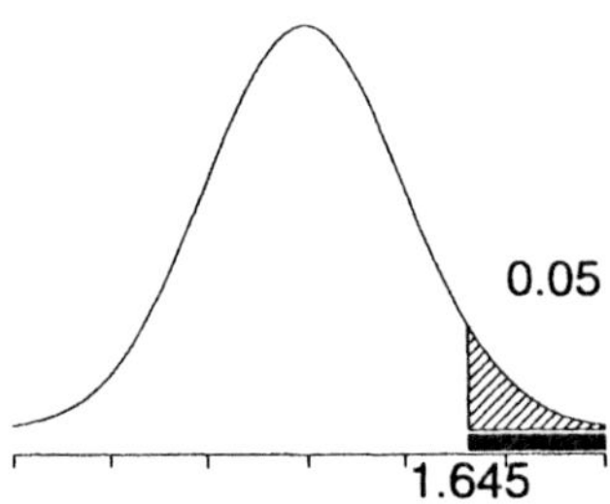

4. *Calculations:* In this case, $x_1 = 57$, $n_1 = 150$, $x_2 = 33$, $n_2 = 100$, and

$$\hat{p} = \frac{57 + 33}{150 + 100} = .36$$

Hence

$$Z = \frac{57/150 - 33/100}{\sqrt{(.36)(.64)(1/150 + 1/100)}} = .807.$$

5. *Decision:* Since the observed value $.807 < z_{.05} = 1.645$, we cannot conclude

at the 5% significance level that the first procedure is better than the second.

9.61 1. *Null hypothesis* $H_0 : p_A = p_B$

Alternative hypothesis $H_1 : p_A < p_B$

2. *Level of significance:* $\alpha = 0.05$.

3. *Criterion:* We using the large sample statistic and reject the null hypothesis when

$$Z = \frac{X_1/n_1 - X_2/n_2}{\sqrt{\hat{p}(1-\hat{p})(\frac{1}{n_1} + \frac{1}{n_2})}} \quad \text{with} \quad \hat{p} = \frac{X_1 + X_2}{n_1 + n_2}$$

is less than $-z_{.05} = -1.645$.

4. *Calculations:* In this case, $x_1 = 11$, $n_1 = 50$, $x_2 = 19$, $n_2 = 50$, and

$$\hat{p} = \frac{11 + 19}{50 + 50} = .30$$

Hence

$$Z = \frac{11/50 - 19/50}{\sqrt{(.30)(.70)(1/50 \ + \ 1/50)}} = -1.75.$$

5. *Decision:* Since the observed value $-1.75 < -z_{.05} = -1.645$, Thus, we reject the null hypothesis at the 5 % level and conclude that agent A is better than agent B.

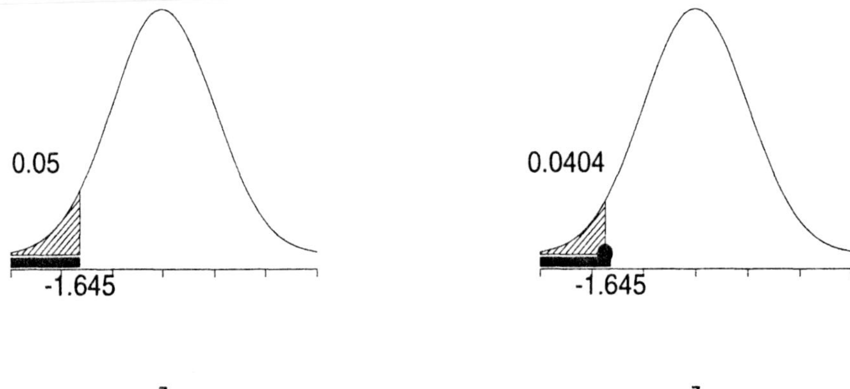

9.63 (a) We use the χ^2 statistic with 2 degrees of freedom to test the null hypothesis that the proportions of clogged cooling pipes at the three plants are equal against the alternative that they are not equal. Thus, we reject the null hypothesis at the 5% level when $\chi^2 > \chi^2_{.05} = 5.991$. In Table 9.8, the expected frequency and the contribution to the χ^2 statistic of each cell are given in the parentheses and brackets respectively.

Table 9.8. Exercise 9.63.

	Plant I	Plant II	Plant III	Total
Clogged	13 (13.33) [.01]	8 (13.33) [2.13]	19 (13.33) [2.41]	40
Unclogged	17 (16.67) [.01]	22 (16.67) [1.71]	11 (16.67) [1.93]	50
Total	30	30	30	90

The χ^2 statistic is

$$\chi^2 = \frac{.33^2}{13.33} + \frac{5.33^2}{13.33} + \frac{5.67^2}{13.33} + \frac{.33^2}{16.67} + \frac{5.33^2}{16.67} + \frac{5.67^2}{16.67} = 8.190.$$

Thus, we reject the null hypothesis.

(b) The 95% confidence interval for the proportion of clogged pipes at the first plant is

$$\frac{13}{30} - 1.96\sqrt{\frac{(13/30)(17/30)}{30}} < p_1 < \frac{13}{30} + 1.96\sqrt{\frac{(13/30)(17/30)}{30}}$$

or

$$.256 < p_1 < .611.$$

At the second plant, the 95% confidence interval for the proportion of clogged pipes is

$$\frac{8}{30} - 1.96\sqrt{\frac{(8/30)(22/30)}{30}} < p_2 < \frac{8}{30} + 1.96\sqrt{\frac{(8/30)(22/30)}{30}}$$

or

$$.108 < p_2 < .425.$$

At the third plant, the 95% confidence interval for the proportion of clogged pipes is

$$\frac{19}{30} - 1.96\sqrt{\frac{(19/30)(11/30)}{30}} < p_3 < \frac{19}{30} + 1.96\sqrt{\frac{(19/30)(11/30)}{30}}$$

or

$$.461 < p_3 < .806.$$

The plot of the confidence intervals is given in Figure 9.1.

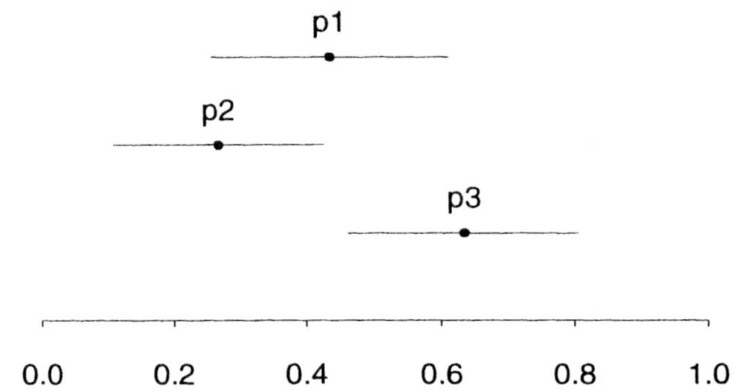

Figure 9.1: Confidence intervals. Exercise 9.63.

9.65 The data are summarized in the table:

Table 9.9. Exercise 9.65.

Number of failures	Number of days		Poisson prob. for $\lambda = 3.2$	Expected frequency	
0	9		.041	12.3	
1	43		.130	39.0	
2	64		.209	62.7	
3	62		.223	66.9	
4	42		.178	53.4	
5	36		.114	34.2	
6	22		.060	18.0	
7	14		.028	8.4	
8	6		.011	3.3	
9	2	8	.004	1.2	5.1
$\geq$ 10	0		.002	.6	

We use the χ^2 statistic with $9-1=8$ degrees of freedom to test the null hypothesis that the data are from a Poisson distribution with $\lambda = 3.2$. We reject the null hypothesis at the 5% level when $\chi^2 > \chi^2_{.05} = 15.507$. In this case,

$$\begin{aligned}\chi^2 &= \frac{3.3^2}{12.3} + \frac{4.0^2}{39.0} + \frac{1.3^2}{62.7} + \\ &\quad \frac{4.9^2}{66.9} + \frac{11.4^2}{53.4} + \frac{1.8^2}{34.2} + \\ &\quad \frac{4.0^2}{18.0} + \frac{5.6^2}{8.4} + \frac{2.9^2}{5.1} \\ &= 10.481.\end{aligned}$$

Thus, we cannot reject the null hypothesis.

9.67 We use the χ^2 statistic with 4 degrees of freedom to test the null hypothesis that appearance and X-ray inspection are independent. Thus, we reject the null hypothesis at the 5% level when $\chi^2 > \chi^2_{.05} = 9.488$. In Table 9.10, the expected frequency and the contribution to the χ^2 statistic of each cell are given in the parentheses and brackets respectively.

The χ^2 statistic is

$$\begin{aligned}\chi^2 &= \frac{12^2}{8} + \frac{7^2}{14} + \frac{5^2}{8} + \\ &\quad \frac{8.33^2}{21.33} + \frac{13.67^2}{37.33} + \frac{5.33^2}{21.33} + \\ &\quad \frac{3.67^2}{10.67} + \frac{6.67^2}{18.67} + \frac{10.33^2}{10.67} \\ &= 47.862.\end{aligned}$$

Thus, we reject the null hypothesis at the 5% level and conclude that there is dependence between appearance and the result of an x-ray inspection. The largest contributions to χ^2 come from the high counts in the Bad-Bad and Good-Good cells.

Table 9.10. Exercise 9.67.

		Appearance			
		Bad	Normal	Good	Total
X-ray	Bad	20 (8) [18.00]	7 (14) [3.50]	3 (8) [3.12]	30
	Normal	13 (21.33) [1.26]	51 (37.33) [5.00]	16 (21.33) [1.33]	80
	Good	7 (10.67) [1.26]	12 (18.67) [2.38]	21 (10.67) [10.01]	40
	Total	40	70	40	150

Chapter 10

NONPARAMETRIC STATISTICS

10.1 1. *Null hypothesis:* $\tilde{\mu} = 0.55$ $(p = \frac{1}{2})$

Alternative hypothesis: $\tilde{\mu} \neq 0.55$ $(p > \frac{1}{2})$

2. *Level of significance:* $\alpha = 0.05$.

3. *Criterion:* The criterion may be based on the number of plus signs or the number of minus signs. Using the number of plus signs, denoted by x, reject the null hypothesis if either the probability of getting x or more plus signs or the probability of getting x or fewer plus signs is less than or equal to $0.05/2$.

4. *Calculations:* The signs of the differences between the observations and $\tilde{\mu} =$ .55 are:

$$+ + - 0 + 0 - + + - + + - + + - + -$$

Ignoring the two 0's, for the cases where the values equal 0.55, we have x =10 and the effective sample size is n =16. From the binomial distribution, Table 1, the probability of $X \geq 10$ with $p = 0.5$ is $1 - .7728 = .2272$. The probability of $X \leq 10$ is .8949.

5. *Decision:* Since both of the probabilities are greater than $0.05/2 = .025$, we fail to reject the null hypothesis $\tilde{\mu} = 0.55$, at the level $\alpha = .05$.

10.3

1. *Null hypothesis:* $\tilde{\mu} = 158$ $(p = \frac{1}{2})$
   Alternative hypothesis: $\tilde{\mu} \neq 158$ $(p \neq \frac{1}{2})$
2. *Level of significance:* $\alpha = 0.05$.
3. *Criterion:* The criterion may be based on the number of plus signs or the number of minus signs. Using the number of plus signs, denoted by x, reject the null hypothesis if either the probability of getting x or more plus signs or the probability of getting x or fewer positives is less than or equal to $0.05/2$.
4. *Calculations:* The signs of the differences between the observations and $\tilde{\mu} = 158$ are:

   $$+ \; - \; - \; - \; + \; + \; + \; - \; - \; - \; + \; - \; - \; - \; + \; + \; -$$

   Since none of the observation are exactly equal 158, the sample size is $n = 15$ and we have $x = 6$ positives. From the binomial distribution, Table 1, the probability of $X \geq 6$, when $p = .5$, is $1 - .1509 = .8491$ and the probability of $X \leq 6$, when $p = .5$, is .3036.
5. *Decision:* Since both of the probabilities are greater than .025, we fail to reject the null hypothesis $\tilde{\mu} = 158$, at the level $\alpha = .05$.

10.5

1. *Null hypothesis:* $\tilde{\mu} = 6.5$ $(p = \frac{1}{2})$
   Alternative hypothesis: $\tilde{\mu} < 6.5$ $(p < \frac{1}{2})$
2. *Level of significance:* $\alpha = 0.01$.
3. *Criterion:* The criterion may be based on the number of plus signs or the number of minus signs. Using the number of plus signs, denoted by x, reject

the null hypothesis if the probability of getting x or fewer positives is less than or equal to 0.01. Using the normal approximation, we reject H_0 if

$$z = \frac{x + .5 - n/2}{\sqrt{n(1/2)(1/2)}} < -2.327$$

4. *Calculations:* The signs of the differences between the observations and $\tilde{\mu} =$ 150 are:

− − − − − − − − − − − − − − − − − − − −

+ − − − + − + + − − − + − + − − + + − −

+ + − + − − − − + − − + + + + − − − − −

− − − − − − + − + − − − − − + − + − + +

The sample size is $n = 80$ and we have $x = 22$ positives. Using the normal approximation to the binomial with $p = 0.5$,

$$z = \frac{22.5 - 80/2}{\sqrt{80(1/2)(1 - 1/2)}} = -3.91$$

5. *Decision:* Since $z = -3.91$, we reject the null hypothesis in favor of $\tilde{\mu} < 6.5$, at the level $\alpha = .01$.

10.7 1. *Null hypothesis:* $\tilde{\mu_D} = 0 (p = \frac{1}{2})$
Alternative hypothesis: $\tilde{\mu_D} \neq 0 \; (p \neq \frac{1}{2})$

2. *Level of significance:* $\alpha = 0.10$.

3. *Criterion:* The criterion may be based on the number of plus signs or the number of minus signs. Using the number of plus signs, denoted by x, reject the null hypothesis if either the probability of getting x or more positives or the probability of getting x or fewer positives is less than or equal to $0.10/2$.

4. *Calculations:* The signs of the differences are:

$$- \quad - \quad - \quad - \quad + \quad - \quad - \quad + \quad - \quad +$$

The sample size is $n = 10$ and we have $x = 3$ positives. From the binomial distribution, Table 1, the probability of $X \leq 3$, when $p = .5$, is .1719 and the probability of 3 or more , when $p = .5$, is $1 - .0547 = .9453$.

5. *Decision:* Since both of these probabilities is greater than .05, we cannot reject the null hypothesis at the 10 percent level.

10.9 1. *Null hypothesis:* Populations are identical.

Alternative hypothesis: Populations are not identical.

2. *Level of significance:* $\alpha = 0.05$.

3. *Criterion:* We reject the null hypothesis if U_1 is too small or too large. That is, we reject H_0 if

$$Z = \frac{U_1 - \mu_{U_1}}{\sigma_{U_1}} < -1.96 \quad or \quad Z > 1.96.$$

4. *Calculations:* The ranks for the first sample are

$$3, 25, 6, 223, 14, 20, 19, 11, 27, 16, 5, 10, 9, 2, 18$$

and the sum of ranks of the first sample is $W_1 = 208$. Then,

$$U_1 = W_1 - \frac{n_1(n_1+1)}{2} = 208 - \frac{15 \cdot 16}{2} = 88.$$

Under the null hypothesis, the mean and variance of the U_1 statistic are

$$\mu_{U_1} = \frac{n_1 \cdot n_2}{2} = \frac{15 \cdot 12}{2} = 90.$$

and

$$\sigma^2_{U_1} = \frac{n_1 \cdot n_2(n_1 + n_2 + 1)}{12} = \frac{15 \cdot 12(15 + 12 + 1)}{12} = 420.$$

Thus, the Z statistic is

$$Z = \frac{88 - 90}{\sqrt{420}} = -0.0976$$

5. *Decision:* We cannot reject the null hypothesis at the .05 level of significance.

10.11 1. *Null hypothesis:* Populations are identical.

Alternative hypothesis: Populations are not identical.

2. *Level of significance:* $\alpha = 0.01$.

3. *Criterion:* We reject the null hypothesis if U_1 is too small or too large. That is, we reject H_0 if

$$Z = \frac{U_1 - \mu_{U_1}}{\sigma_{U_1}} < -2.575 \quad or \quad Z > 2.575.$$

4. *Calculations:* The ranks for the first sample are

$$16, 21, , 5, 14.5, 13, 12, 18, 22, 20, 23, 17, 19$$

The sum of ranks of the first sample is

$$W_1 = 200.5$$

Thus,

$$U_1 = W_1 - \frac{n_1(n_1 + 1)}{2} = 200.5 - \frac{12 \cdot 13}{2} = 122.5.$$

Under the null hypothesis, the mean and variance of the U_1 statistic are

$$\mu_{U_1} = \frac{n_1 \cdot n_2}{2} = \frac{12 \cdot 12}{2} = 72.$$

and

$$\sigma^2_{U_1} = \frac{n_1 \cdot n_2(n_1 + n_2 + 1)}{12} = \frac{12 \cdot 12(12 + 12 + 1)}{12} = 300.$$

Thus, the Z statistic is

$$Z = \frac{122.5 - 72}{\sqrt{300}} = 2.916$$

5. *Decision:* We reject the null hypothesis in favor of the alternative that the populations are not identical, at the .01 level of significance.

10.13 1. *Null hypothesis:* Populations are identical.

Alternative hypothesis: The first population is stochastically larger than the second.

2. *Level of significance:* $\alpha = 0.05$.

3. *Criterion:* We reject the null hypothesis if U_1 is too large. That is, we reject H_0 if

$$Z = \frac{U_1 - \mu_{U_1}}{\sigma_{U_1}} > 1.645$$

4. *Calculations:* The ranks for the first sample are

$$3, 20, 32, 25, 11.5, 14, 24, 27.5, 17.5, 21, 1, 22, 29, 10, 19, 30.5$$

The sum of ranks of the first sample is

$$W_1 = 307$$

Thus,

$$U_1 = W_1 - \frac{n_1(n_1+1)}{2} = 307 - \frac{16 \cdot 17}{2} = 171.$$

Under the null hypothesis, the mean and variance of the U_1 statistic are

$$\mu_{U_1} = \frac{n_1 \cdot n_2}{2} = \frac{16 \cdot 16}{2} = 128$$

and

$$\sigma^2_{U_1} = \frac{n_1 \cdot n_2(n_1 + n_2 + 1)}{12} = \frac{16 \cdot 16(16 + 16 + 1)}{12} = 704.$$

Thus, the Z statistic is

$$Z = \frac{171 - 120}{\sqrt{704}} = 1.621$$

5. *Decision:* We cannot reject the null hypothesis at the .05 level of significance. In other words , we cannot conclude that strength of material 1 is stochastically larger than that of material 2.

10.15 1. *Null hypothesis:* The populations are identical.

Alternative hypothesis: Populations are not all equal.

2. *Level of significance:* $\alpha = 0.05$.

3. *Criterion:* We reject the null hypothesis if $H > 9.488$ the value of $\chi^2_{.05}$ for 4 degrees of freedom.

4. *Calculations:* The sums of ranks for the samples are

$$\begin{aligned}
R_1 &= 20.5 + 22 + 23.5 + 23.5 + 26.5 + 31 = 147.0 \\
R_2 &= 5 + 12 + 13.5 + 13.5 + 15 + 17 + 20.5 = 96.5 \\
R_3 &= 1 + 2 + 3 + 7 + 7 + 10.5 = 30.5 \\
R_4 &= 19 + 25 + 26.5 + 28 + 29 + 30 + 32 + 33 = 222.5 \\
R_5 &= 4 + 7 + 9 + 10.5 + 16 + 18 = 64.5
\end{aligned}$$

Thus,

$$H = \frac{12}{30 \cdot 31} [\frac{(147)^2}{6} + \frac{(96.5)^2}{7} + \frac{(30.5)^2}{6}$$
$$+ \frac{(222.5)^2}{8} + \frac{(64.5)^2}{6}] - 3 \cdot 34 = 26.01.$$

5. *Decision:* We reject the null hypothesis at the .05 level of significance. The P-value is much smaller than .005.

10.17

1. *Null hypothesis:* Arrangement is random.

   Alternative hypothesis: Arrangement is not random.

2. *Level of significance:* $\alpha = 0.05$.

3. *Criterion:* We reject the null hypothesis if

$$Z = \frac{u - \mu_u}{\sigma_u} < -1.96 \quad or \quad Z > 1.96.$$

where u is the total number of runs and

$$\mu_u = \frac{2n_1n_2}{n_1 + n_2} + 1$$

and

$$\sigma_u^2 = \frac{2n_1n_2(2n_1n_2 - n_1 - n_2)}{(n_1 + n_2)^2(n_1 + n_2 - 1)}$$

4. *Calculations:* The runs in the data are underlined.

$\underline{LL}\ \underline{O}\ \underline{LLLL}\ \underline{OO}\ \underline{LLLL}\ \underline{O}\ \underline{L}\ \underline{OO}\ \underline{LLLL}\ \underline{O}\ \underline{L}\ \underline{OO}\ \underline{LLLLL}$

$\underline{O}\ \underline{LLL}\ \underline{OL}\ \underline{O}\ \underline{LLLL}\ \underline{OO}\ \underline{L}\ \underline{OOOO}\ \underline{LLLL}\ \underline{O}\ \underline{L}\ \underline{OO}\ \underline{LLL}\ \underline{O}$

There are $u = 28$ runs, 22 O's and 38 L's. Under the null hypothesis, the mean and standard deviation are

$$\mu_u = \frac{2n_1n_2}{n_1+n_2}+1 = \frac{2\cdot 22\cdot 38}{22+38}+1 = 28.87$$

$$\sigma_u = \sqrt{\frac{2n_1n_2(2n_1n_2-n_1-n_2)}{(n_1+n_2)^2(n_1+n_2-1)}}$$

$$= \sqrt{\frac{2\cdot 22\cdot 38(2\cdot 22\cdot 38-22-38)}{(22+38)^2(22+38-1)}} = 3.562$$

Thus, the Z statistic is

$$Z = \frac{28-28.87}{3.562} = -.244$$

5. *Decision:* We cannot reject the null hypothesis at the .05 level of significance.

10.19

1. *Null hypothesis:* Arrangement is random.

   Alternative hypothesis: Arrangement is not random.

2. *Level of significance:* $\alpha = 0.01$.

3. *Criterion:* We reject the null hypothesis if

$$Z = \frac{u-\mu_u}{\sigma_u} < -2.575 \quad or \quad Z > 2.575.$$

   where u is the total number of runs and

$$\mu_u = \frac{2n_1n_2}{n_1+n_2}+1$$

   and

$$\sigma_u^2 = \frac{2n_1n_2(2n_1n_2-n_1-n_2)}{(n_1+n_2)^2(n_1+n_2-1)}$$

4. *Calculations:* The sequence of odds and evens in the first 5 rows of Table 7 is

OOEE OOEO OOOO OOEE OEEE

$$OEEE\ EOEO\ OEOO\ OEOE\ OEOE$$

$$EEEE\ EEOO\ EEEO\ EEEO\ OEOO$$

$$EOEE\ EOOE\ OEEE\ OEEE\ OOEO$$

$$EOOO\ EEEE\ EOEE\ OEEE\ EEOE$$

$$EOOO\ EOEE\ OEEO\ OEOO\ EOEE$$

$$OOEO\ EOEO\ OEEO\ EEEO\ OEOE$$

$$EOEO\ OOOO\ OOEE\ OEEE\ EOEE$$

$$OEOO\ EOOO\ EEEE\ EEOE\ EEEO$$

$$EOOO\ OOOO\ OEEO\ OOEO\ OOOO$$

There are $u = 99$ runs, the number of odds is 97, and the number of evens is 103. Under the null hypothesis, the mean and standard deviation are

$$\begin{aligned} \mu_u &= \frac{2n_1n_2}{n_1+n_2} + 1 = \frac{2\cdot 97\cdot 103}{97+103} + 1 = 100.91 \\ \sigma_u &= \sqrt{\frac{2n_1n_2(2n_1n_2 - n_1 - n_2)}{(n_1+n_2)^2(n_1+n_2-1)}} \\ &= \sqrt{\frac{2\cdot 97\cdot 103)(2\cdot 97\cdot 103 - 97 - 103)}{(97+103)^2(97+103-1)}} = 7.0469 \end{aligned}$$

Thus, the Z statistic is

$$Z = \frac{99 - 100.91}{7.0469} = -.2710$$

5. *Decision:* We cannot reject the null hypothesis of randomness at the .01 level of significance or at any reasonable level.

10.21 1. *Null hypothesis:* The arrangement of sample values is random.

Alternative hypothesis: The arrangement is not random.

2. *Level of significance:* $\alpha = 0.05$.

3. *Criterion:* We reject the null hypothesis if

$$Z = \frac{u - \mu_u}{\sigma_u} < -1.96 \quad or \quad Z > 1.96.$$

where u is the total number of runs above and below the median.

4. *Calculations:* The median of the data is 36. Using a for observations above the median and b for those below, the data are:

$$\underline{bbbbbb}\ \underline{a}\ \underline{bbbbbbb}\ \underline{aa}\ \underline{bbbbbbb}\ \underline{aaa}\ \underline{bbb}\ \underline{aa}\ \underline{b}\ \underline{aaaaa}\ \underline{b}\ \underline{aaaaaaaaaaa}$$

There are $u = 12$ runs, 25 a's and 25 b's. In this case

$$\mu_u = \frac{2 \cdot 25 \cdot 25}{25 + 25} + 1 = 26$$

and

$$\sigma_u = \sqrt{\frac{2 \cdot 25 \cdot 25(2 \cdot 25 \cdot 25 - 25 - 25}{25 + 25)^2(25 + 25 - 1)}} = 3.448$$

so

$$Z = \frac{12 - 26}{3.449} = -4.005$$

5. *Decision:* We reject the null hypothesis. at the .05 level of significance. The P- value is .00006.

10.23 The null hypothesis is that the data are from an exponential distribution with mean 10. The alternative is that the data are not from this exponential distribution. to test the null hypothesis against the alternative, at the 5 percent level, use the statistic D and reject when $D > .309$ since $n = 18$. Since D is the maximum absolute difference between the empirical distribution and the exponential distribution and , according to Figure 10.1 D is approximately .26, we cannot reject the null hypothesis. In general, much larger sample sizes are required to detect lack of fit.

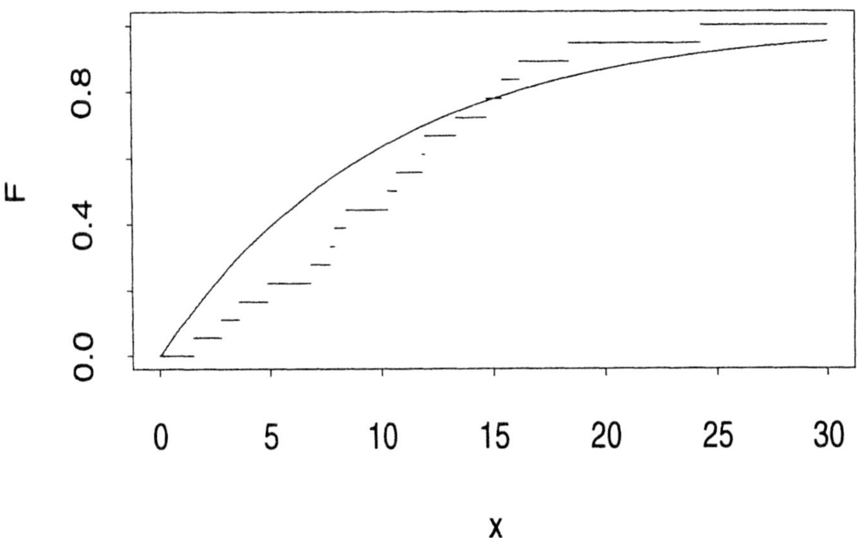

Figure 10.1: Empirical and null distributions. Exercise 10.23.

10.25

1. *Null hypothesis:* $\tilde{\mu} = 0 (p = \frac{1}{2})$

   Alternative hypothesis: $\tilde{\mu} > 0$ $(p > \frac{1}{2})$

2. *Level of significance:* $\alpha = 0.063$.

3. *Criterion:* The criterion may be based on the number of plus signs or the number of minus signs. Using the number of plus signs, denoted by x, reject the null hypothesis if the probability of getting x or more positives is less than or equal to 0.063.

4. *Calculations:* The signs of the differences are:

$$+ + - + + + + +$$

The sample size is $n = 7$ and we have $x = 6$ positives. From the binomial distribution, Table 1, the probability of $X \geq 6$, when $p = .5$, is $1 - .9357 = .0625$.

5. *Decision:* Since the probability .0625 is less than .063 we reject the null hypothesis.

10.27

1. *Null hypothesis:* Populations are identical.

   Alternative hypothesis: The Method B population is stochastically larger than the Method A population.

2. *Level of significance:* $\alpha = 0.05$.

3. *Criterion:* We reject the null hypothesis if U_1 is too small. That is, we reject H_0 if

$$Z = \frac{U_1 - \mu_{U_1}}{\sigma_{U_1}} < -1.645.$$

4. *Calculations:* The ranks for the first sample are

$$1.5, 3, 4.5, 4.5, 6.5, 9.5, 12.5, 14, 15.5$$

The sum of ranks of the first sample is

$$W_1 = 1.5 + 3 + 4.5 + 4.5 + 6.5 + 9.5 + 12.5 + 14 + 15.5 = 81$$

Thus ,

$$U_1 = W_1 - \frac{n_1(n_1+1)}{2} = 81 - \frac{10 \cdot 11}{2} = 26.$$

Under the null hypothesis, the mean and variance of the U_1 statistic are

$$\mu_{U_1} = \frac{n_1 \cdot n_2}{2} = \frac{10 \cdot 10}{2} = 50.$$

and

$$\sigma^2_{U_1} = \frac{n_1 \cdot n_2(n_1 + n_2 + 1)}{12} = \frac{10 \cdot 10(10 + 10 + 1)}{12} = 175.$$

Thus, the Z statistic is

$$Z = \frac{26 - 50}{\sqrt{175}} = -1.814$$

5. *Decision:* We reject the null hypothesis at the .05 level of significance and conclude that the Method B population is stochastically larger than the Method A population.

10.29 1. *Null hypothesis:* The three populations are identical.

Alternative hypothesis: Populations are not all equal.

2. *Level of significance:* $\alpha = 0.05$.

3. *Criterion:* We reject the null hypothesis if $H > 5.991$ the value of $\chi^2_{.05}$ for 2 degrees of freedom.

4. *Calculations:* The sums of ranks for the samples are

$$R_1 = 1.5 + 5 + 7.5 + 10.5 + 12 + 13 + 15.5 + 18 + 25 + 28 = 136$$

$$R_2 = 3 + 5 + 7.5 + 9 + 10.5 + 20 + 21 + 22.5 + 28 + 30 = 156.5$$

$$R_3 = 1.5 + 5 + 14 + 15.5 + 18 + 18 + 22.5 + 25 + 25 + 28 = 172.5$$

Thus ,

$$H = \frac{12}{30 \cdot 31}\left[\frac{(136)^2}{10} + \frac{(156.5)^2}{10} + \frac{(172.5)^2}{10}\right] - 3 \cdot 31 = 0.904.$$

5. *Decision:* We cannot reject the null hypothesis at the .05 level of significance.

10.31 1. *Null hypothesis:* The arrangement of sample values is random.

Alternative hypothesis: There is a trend in the sample.

2. *Level of significance:* $\alpha = 0.05$.

3. *Criterion:* We reject the null hypothesis if

$$Z = \frac{u - \mu_u}{\sigma_u} < -1.645.$$

where u is the total number of runs above and below the median.

4. *Calculations:* The median of the data is 138. Using a for observations above the median and b for those below, the data are:

$$\underline{bbbbb}\ \underline{aa}\ \underline{bb}\ \underline{aaaa}\ \underline{b}\ \underline{a}\ \underline{bbb}\ \underline{aaaa}\ \underline{bbb}\ \underline{a}\ \underline{bb}\ \underline{s}\ \underline{aaaa}$$

Ignoring the tie (s), there are $u = 12$ runs, 16 a's and 16 b's. In this case

$$\mu_u = \frac{2 \cdot 16 \cdot 16}{16 + 16} + 1 = 17$$

and

$$\sigma_u = \sqrt{\frac{2 \cdot 16 \cdot 16(2 \cdot 16 \cdot 16 - 16 - 16}{16 + 16)^2(16 + 16 - 1)}} = 2.782$$

so

$$Z = \frac{12 - 17}{2.782} = -1.797$$

5. *Decision:* We reject the null hypothesis at the .05 level of significance and conclude that there is a trend over time.

10.33 1. *Null hypothesis:* Populations are identical.

Alternative hypothesis: The Heat 1 population is stochastically larger than the Heat 2 population.

2. *Level of significance:* $\alpha = 0.033$.

3. *Criterion:* We reject the null hypothesis if U_1 is too large. Since the distribution of U_1 is symmetric about $n_1 n_2/2 = 10.5$, $.033 = P(U_1 \leq 2) = P(U_1 \geq 19)$ and we reject H_0 if $U_1 \geq 19$.

4. *Calculations:* The sum of ranks of the first sample is

$$W_1 = 6 + 9 + 10 = 25$$

Thus,

$$U_1 = W_1 - \frac{n_1(n_1+1)}{2} = 25 - \frac{3 \cdot 4}{2} = 19.$$

5. *Decision:* We reject the null hypothesis at the .033 level of significance and conclude that the Method B population is stochastically larger than the Method A population.

10.35 1. *Null hypothesis:* The arrangement of sample values is random.
Alternative hypothesis: The arrangement is not random.

2. *Level of significance:* $\alpha = 0.05$.

3. *Criterion:* We reject the null hypothesis if

$$Z = \frac{u - \mu_u}{\sigma_u} < -1.96 \quad \text{or} \quad Z > 1.96.$$

where u is the total number of runs above and below the median.

4. *Calculations:* The median of the differences is 2. Using a for observations above the median and b for those below, the data are:

$$\underline{s}\ \underline{bbbbb}\ \underline{a}\ \underline{b}\ \underline{s}\ \underline{aaaaaaa}\ \underline{s}\ \underline{bbb}$$

Ignoring the ties (s), there are $u = 5$ runs, $n_1 = 8$ a's and $n_2 = 9$ b's. In this case

$$\mu_u = \frac{2 \cdot 8 \cdot 9}{8+9} + 1 = 9.471$$

and

$$\sigma_u = \sqrt{\frac{2 \cdot 8 \cdot 9(2 \cdot 8 \cdot 9 - 8 - 9)}{(8+9)^2(8+9-1)}} = 1.989$$

so

$$Z = \frac{6 - 9.471}{1.989} = -2.243$$

5. *Decision:* We reject the null hypothesis at the .05 level of significance. There are two few runs suggesting that adjacent positions are positively correlated.

Chapter 11

CURVE FITTING

11.1 (a) Use a random number table to select amount of additive for first run. Ignore 0, 6, 7, 8, and 9. Do the same for each of the four other runs but ignore amounts already selected.

(b) The experiment used the values 1, 2, 3, 4, and 5 but 8 is quite far beyond this range of values. Don't extrapolate beyond the range of experimental values because the model may change.

11.3 (a) The hand-drawn line in the scattergram, Figure 11.1, yields a prediction of 67 percent for the extraction efficiency when the extraction time is 35 minutes.

(b) In this example, n=10

$$\sum_{i=1}^{n} y_i = 635, \quad \sum_{i=1}^{n} x_i = 320, \quad \sum_{i=1}^{n} x_i^2 = 11,490, \quad \sum_{i=1}^{n} x_i y_i = 21,275$$

so,

$$\begin{aligned} S_{xx} &= \sum_{i=1}^{n} x_i^2 - (\sum_{i=1}^{n} x_i)^2/n = 11,490 - (320)^2/10 = 1250, \\ S_{xy} &= \sum_{i=1}^{n} x_i y_i - (\sum_{i=1}^{n} x_i)(\sum_{i=1}^{n} y_i)/n = 21,275 - 320(635)/10 = 955 \end{aligned}$$

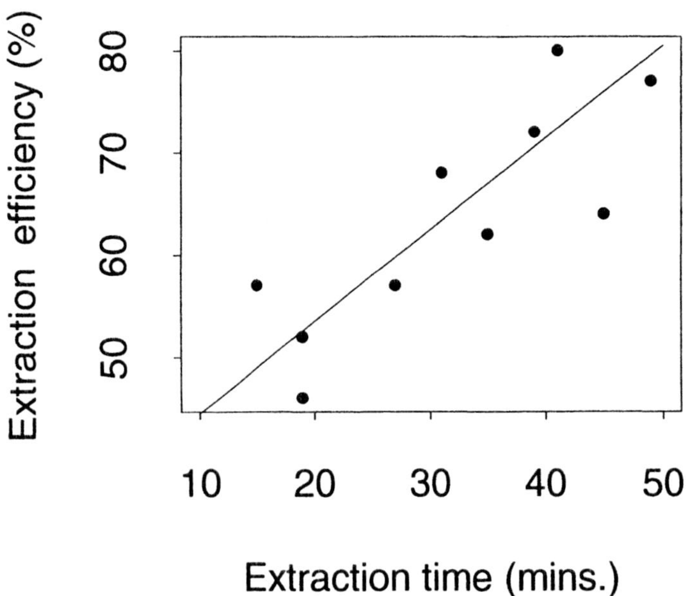

Figure 11.1: Scattergram for Exercise 11.3.

Consequently,

$$
\begin{aligned}
b &= \frac{S_{xy}}{S_{xx}} = \frac{955}{1250} = .764 \\
a &= \overline{y} - b\overline{x} = \frac{635}{10} - \frac{955}{1250}\,\frac{320}{10} = 39.052
\end{aligned}
$$

Thus, the equation for the least squares line is:

$$y = 39.052 + .764x$$

The prediction of the extraction efficiency when the extraction time is 35 minutes is

$$39.052 + (.764)(35) = 65.79 \text{ percent}$$

11.5 (a) Using the sums from Exercise 11.4b,

$$\begin{aligned} S_{xx} &= \sum_{i=1}^{n} x_i^2 - (\sum_{i=1}^{n} x_i)^2/n = 91 - (21)^2/6 = 17.5, \\ S_{yy} &= \sum_{i=1}^{n} y_i^2 - (\sum_{i=1}^{n} y_i)^2/n = 19855 - (311)^2/6 = 3,734.834, \\ S_{xy} &= \sum_{i=1}^{n} x_i y_i - (\sum_{i=1}^{n} x_i)(\sum_{i=1}^{n} y_i)/n = 1342 - 21(311)/6 = 253.5 \end{aligned}$$

Thus,

$$s_e^2 = \frac{S_{xx}S_{yy} - (S_{xy})^2}{(n-2)S_{xx}} = \frac{(17.5)(3,734.834) - (253.5)^2}{(4)(17.5)} = 15.676.$$

Thus, the 95 percent confidence interval is given by

$$b \pm t_{\frac{\alpha}{2}} \, s_e \sqrt{\frac{1}{S_{xx}}}$$

or, since $t_{.025}$ for 4 degrees of freedom is 2.776,

$$14.486 \pm 2.776\sqrt{15.676}\sqrt{\frac{1}{17.5}}.$$

So the interval is from 11.86 to 17.11 thousandths of an inch per thousand pounds.

(b) The limits of prediction at x_0 are

$$a + bx_0 \pm t_{\frac{\alpha}{2}} \, s_e \sqrt{1 + \frac{1}{n} + \frac{(x_0 - \bar{x})^2}{S_{xx}}}$$

For Exercise 11.4, when the tensile force is 3.5 thousand pounds, this becomes

$$51.83 \pm 2.776\sqrt{15.676}\sqrt{1+\frac{1}{6}+\frac{(3.5-3.5)^2}{17.5}}$$

Thus, the interval is from 39.96 to 63.71 thousandths of an inch.

11.7 1. *Null hypothesis:* $\beta = 1.2$

Alternative hypothesis: $\beta < 1.2$.

2. *Level of significance:* $\alpha = 0.05$.

3. *Criterion:* Reject the null hypothesis if $t < -2.132$, where 2.132 is the value of $t_{0.05}$ for $6 - 2 = 4$ degrees of freedom , and t is given by the second formula of Theorem 11.1.

4. *Calculations:* We must first calculate S_{xx}, S_{xy}, S_{yy} and s_e. Since $\sum y_i^2 = 2001$, using the results of Exercise 11.4(a) gives

$$\begin{aligned} S_{xx} &= 304 - (36)^2/6 = 88 \\ S_{xy} &= 721 - (36)(107)/6 = 79 \\ S_{yy} &= 2001 - (107)^2/6 = 92.833 \end{aligned}$$

so that

$$s_e^2 = \frac{(88)(92.833) - (79)^2}{(4)(88)} = 5.478$$

Also, from Exercise 11.4(a), $b = .8977$ and $\beta = 1.2$ under the null hypothesis, and hence,

$$t = \frac{b-\beta}{s_e}\sqrt{S_{xx}} = \frac{.8977-1.2}{\sqrt{5.478}}\sqrt{88} = -1.212$$

5. *Decision:* Since -1.212 is greater than -2.132, we fail to reject the null hypothesis at the 0.05 level of significance.

11.9 (a) We calculate

	x	y	$x-\bar{x}$	$y-\bar{y}$	$(x-\bar{x})(y-\bar{y})$	$(x-\bar{x})^2$	$(y-\bar{y})^2$
	1	2	-2	-4	8	4	16
	2	5	-1	-1	1	1	1
	3	4	0	-2	0	0	4
	4	9	1	3	3	1	9
	5	10	2	4	8	4	16
Total	15	30	0	0	20	10	46

So $S_{xx} = 10$, $S_{yy} = 46$ and $S_{xy} = 20$. Also, $\bar{x} = 3$ and $\bar{y} = 6$. Thus,

$$b = \frac{100}{50} = 2$$
$$a = 6 - (2)(3) = 0$$

So, the least squares line is

$$\hat{y} = 0 + 2x$$

(b) When $x = 3.5$, the prediction is

$$\hat{y} = 2(3.5) = 7$$

11.11 We have to test the null hypothesis $H_0: \beta = 1$ against the alternative $H_1: \beta > 1$ at 0.05 level of significance. The t statistic is given by

$$t = \frac{b - \beta}{s_e}\sqrt{S_{xx}}$$

In our case, $S_{xx} = 10$ and $b = 2$. So,

$$t = \frac{(2 - 1)}{\sqrt{2}}\sqrt{10} = 2.236$$

Since $t_{.05}$ is 2.353 for 3 degrees of freedom, we cannot reject the null hypothesis

$H_0: \ \beta = 1.$

11.13 This calculation is the same as in 11.10b except that 1 is added under the square root and t is now $t_{.025}$ with 10 degrees of freedom. Thus, the interval for prediction is given by

$$9.812 \pm 2.228\ (1.101)\ \sqrt{1 + \frac{1}{12} + \frac{(40 - 44.4)^2}{854.9167}}$$

So the interval for the moisture content prediction is from 7.232 to 12.392.

11.15 $S_{xx} = 44,\ S_{xy} = 1097$ from the previous problem.

$$S_{yy} = \sum_{i=1}^{8} y_i^2 - (\sum_{i=1}^{8} y_i)^2/8 \ = \ 27830.25.$$

Thus,

$$s_e^2 \ = \ \frac{S_{xx}S_{yy} - (S_{xy})^2}{6S_{xx}} \ = \ 80.0076 \quad \text{so} \quad s_e \ = \ 8.945$$

With 6 degrees of freedom, $t_{.05} = 1.943$, so the interval is given by

$$24.93 \pm 1.943\ (8.945)\ \sqrt{\frac{1}{44}}.$$

The interval is from 22.31 to 27.55 .

11.17 (a) We wish to minimize

$$\begin{aligned}\sum_{i=1}^{n}(y_i - \beta x_i)^2 \ &= \ \sum_{i=1}^{n}(y_i - bx_i + bx_i - \beta x_i)^2 \\ &= \sum_{i=1}^{n}(y_i - bx_i)^2 + (b-\beta)^2\sum_{i=1}^{n} x_i^2 + 2(\beta - b)\sum_{i=1}^{n}(y_i - bx_i)x_i \\ &= \sum_{i=1}^{n}(y_i - bx_i)^2 + (b-\beta)^2\sum_{i=1}^{n} x_i^2 \ + \ 0\end{aligned}$$

by the definition of b. The first term does not depend on β and the second is a minimum for $\beta = b$, thus establishing the result.

(b) In Exercise 11.2, $\sum_{i=1}^{n} y_i x_i = 1342$ and $\sum_{i=1}^{n} x_i^2 = 91$, so that $b = 1342/91 =$

14.747. The previous slope estimate was 14.486. This estimate (14.747) is a little larger.

(c) In Exercise 11.12, $\sum_{i=1}^{n} y_i x_i = 6217$ and $\sum_{i=1}^{n} x_i^2 = 244$, so that $b = 6217/244$ $= 25.4795$. The previous slope estimate was 24.93. This estimate (25.4795) is a little larger again.

11.19 To calculate s_e^2, we need S_{yy}, which is equal to 21,401,588. Thus,

$$\begin{aligned} s_e^2 &= \frac{(215,065.88)(21,401,588) - (2,145,358.8)^2}{(6)(215,065.88)} \\ &= 144.89816, \end{aligned}$$

or

$$s_e = 12.0373.$$

Thus, the 95 percent limits are ($t_{.05}$ for 6 d.f. $= 1.943$)

$$22.90 \pm 1.943\ (12.0373)\sqrt{\frac{1}{8} + \frac{(117.625)^2}{215,065.88}}$$

or from 12.713 to 33.067.

11.21 (a) Instead of $x = (1,2,3,4,5,6,7)$, we will use $x = (-3,-2,-1,0,1,2,3)$. Then, $\bar{x} = 0$. Thus,

$$a = \sum_{i=1}^{n} y_i/n \quad \text{and} \quad b = \sum_{i=1}^{n} x_i y_i / \sum_{i=1}^{n} x_i^2$$

Since, $\sum_{i=1}^{n} y_i = 23.7$, $\sum_{i=1}^{n} x_i y_i = 19$, $\sum_{i=1}^{n} x_i^2 = 28$,

$$a = 23.7/7 = 3.386 \quad \text{and} \quad b = .679$$

Thus, the least squares line is $\hat{y} = 3.386 + .649x$. The 8'th year corresponds to $x = 4$, so the prediction is

$$3.386 + .679(4) = 6.102.$$

(b) We use $x = (-7, -5, -3, -1, 1, 3, 5, 7)$. Thus, $\sum_{i=1}^{n} y_i = 23.6$, $\sum_{i=1}^{n} x_i y_i = 40.6$, $\sum_{i=1}^{n} x_i^2 = 168$, so,

$$a = 23.6/8 = 2.95 \quad \text{and} \quad b = .24167$$

Thus, the least squares line is $\hat{y} = 2.95 + .24167x$. Since 1996 corresponds to $x = 11$, so the prediction for y is

$$\hat{y} = 2.95 + .24167(11) = 5.608$$

11.23 From Exercise 11.9 we form the following table

	x	y	$\hat{y} = 2x$	$(y - \hat{y})^2$	$(\bar{y} - \hat{y})^2$	$(y - \bar{y})^2$
	1	2	2	0	16	16
	2	5	4	1	4	1
	3	4	6	4	0	4
	4	9	8	1	4	9
	5	10	10	0	16	16
Total	15	30	30	6	40	46

Thus, the decomposition of the sum of squares is calculated as $46 = 6 + 40$, and

$$r^2 = 1 - 6/46 = 40/46 = 0.870$$

11.25 (a) See Figure 11.2.

(b) Let $z_i = \log y_i$. Then, $\bar{x} = 10.5$, $\bar{z} = 5.476$, $S_{xx} = 157.5$, $S_{xz} = 9.507$. Thus,

$$b = \frac{S_{xz}}{S_{xx}} = 0.0604 \text{ and } a = \bar{z} - b\bar{x} = 4.842$$

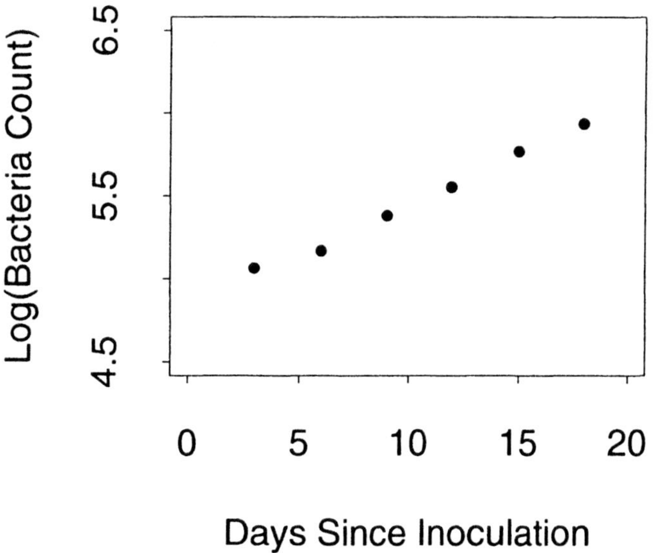

Figure 11.2: Scattergram for Exercise 11.25

So, the fit is

$$\hat{y} = 10^{4.842 + 0.0604x} = 69{,}502.2(1.149)^x$$

(c) Thus, the prediction when $x = 20$, is

$$10^{4.842 + 0.0604(20)} = 1{,}122{,}018$$

11.27 The equation obtained in Exercise 11.26 is

$$\hat{y} = 10^{1.452 + .0000666x} = (10^{1.452})(10^{.0000666x}) = 28.314e^{.0001534x}$$

When $x = 3000$, $\hat{y} = 44.86$.

11.29 The data are

Altitude	Dose Rate
50	28
450	30
780	32
1,200	36
4,400	51
4,800	58
5,300	69

Since the model is

$$y = \exp\left[e^{\alpha x + \beta}\right]$$

we have

$$\ln(\ln y) = \alpha x + \beta$$

Thus, the data to which we are fitting a least squares line are

Altitude	ln (ln (Dose Rate))
50	1.2036
450	1.2241
780	1.2429
1,200	1.2703
4,400	1.3691
4,800	1.4013
5,300	1.4432

Let $z = \ln(\ln y)$, then

$$n = 7, \quad \sum x = 16,980 \quad \sum x^2 = 72,743,400$$

$$\sum z = 9.1605, \quad \sum z^2 = 12.042, \quad \sum xz = 23511.29$$

Hence

$$S_{xx} = 72,743,400 - (16,980)^2/7 = 31,554,771$$

$$S_{xz} = 23511.29 - (16,980)(9.1605)/7 = 1290.5343$$

Consequently, we get

$$b = \frac{S_{xz}}{S_{xx}} = 0.0000409, \quad \text{and} \quad a = \bar{z} - b\bar{x} = 1.2094$$

Thus, the fitted equation is $\hat{y} = \exp(\exp(0.00004x + 1.2))$.

11.31 $y = 3 - 3e^{-\alpha x}$ can be rewritten as

$$\ln(1 - y/3) = -\alpha x$$

Let $y_i^{'} = \ln(1 - y_i/3)$ and $x_i^{'} = -x_i$. Thus, using the no-intercept least square fit

$$\alpha = \frac{\sum x^{'} y^{'}}{\sum x^{'2}} = \frac{264.3584}{1100} = .240$$

11.33 (a) Fitting a straight line to the data gives

$$\hat{y} = 10.4778 - 0.38334x$$

where

$$SSE_1 = \sum (y - \hat{y})^2 = 11.906,$$

thereby giving

$$\hat{\sigma_1}^2 = 11.906/9 = 1.32289.$$

The t statistic for the null hypothesis $\beta_1 = 0$ is given by

$$t = \frac{-0.38334 - 0}{\sqrt{11.906/7}} \sqrt{\frac{9.60}{9}} = -2.28.$$

(b) In this case, $\hat{\sigma_2}^2 = \sum (y - \hat{y})^2/\text{d.f.} = 1.602/6 = 0.2670$ and

$\hat{y} = 12.1848 - 1.8465 + 1.829x^2$. Thus, the F statistic is given by

$$\frac{11.906 - 1.602}{0.2670} = \frac{10.304}{0.2670} = 38.59176$$

The critical value for $F_{.05}$ with 1 and 6 degrees of freedom is 5.99 so the test is not significant at the 5 percent level.

11.35 We wish to minimize

$$D \equiv D(\beta_0, \beta_1, \beta_2) = \sum_{i=1}^{n} [y_i - (\beta_0 + \beta_1 x_{1i} + \beta_2 x_{2i})]^2$$

Taking derivatives

$$\begin{aligned}
\frac{\partial D}{\partial \beta_0} &= -2 \sum_{i=1}^{n} [y_i - (\beta_0 + \beta_1 x_{1i} + \beta_2 x_{2i})] \\
\frac{\partial D}{\partial \beta_1} &= -2 \sum_{i=1}^{n} [y_i - (\beta_0 + \beta_1 x_{1i} + \beta_2 x_{2i})]\, x_{1i} \\
\frac{\partial D}{\partial \beta_2} &= -2 \sum_{i=1}^{n} [y_i - (\beta_0 + \beta_1 x_{1i} + \beta_2 x_{2i})]\, x_{2i}
\end{aligned}$$

Setting these to zero gives

$$\begin{aligned}
\sum y &= nb_0 + b_1 \sum x_1 + b_2 \sum x_2 \\
\sum x_1 y &= b_0 \sum x_1 + b_1 \sum x_1^2 + b_2 \sum x_1 x_2 \\
\sum x_2 y &= b_0 \sum x_2 + b_1 \sum x_1 x_2 + b_3 \sum x_2^2
\end{aligned}$$

11.37 Using $\hat{y} = 161.34 + 32.97x_1 - .086x_2$ gives $\hat{y} = 64.0885$ for $x_1 = .05$ and $x_2 = 1150$.

11.39 The system of linear equations to be solved is

$$84.6 = 10b_0 + 15.3b_1 + 939b_2$$

$$132.27 = 15.3b_0 + 29.85b_1 + 1{,}458.9b_2$$
$$8{,}320.2 = 939b_0 + 1{,}458.9b_1 + 94{,}131b_2$$

Solving gives the fitted line

$$\hat{y} = 2.266 + 0.225x_1 + 0.0623x_2$$

When $x_1 = 2.2$ and $x_2 = 90$, the predicted value is 8.368.

11.41

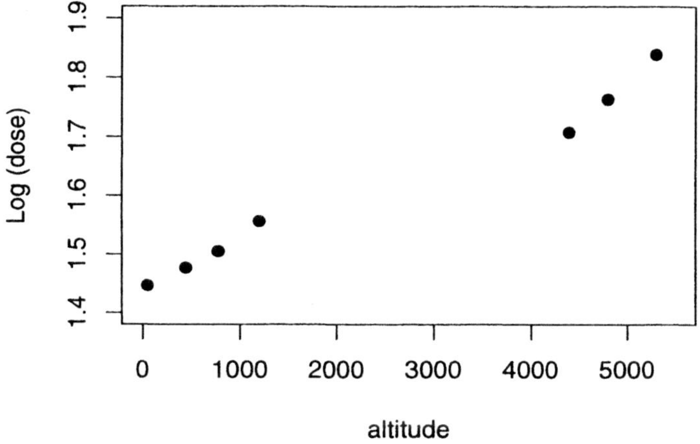


```
The regression equation is
Log(dose) = 1.45 +0.000067 altitude

Predictor         Coef      SE Coef          T        P
Constant       1.45203      0.01475      98.41    0.000
altitude    0.00006663   0.00000458      14.56    0.000

S = 0.02571     R-Sq = 97.7%     R-Sq(adj) = 97.2%

Analysis of Variance

Source          DF         SS          MS         F        P
Regression       1    0.14008     0.14008    211.91    0.000
Residual Error   5    0.00331     0.00066
Total            6    0.14339
```

11.43 (a)

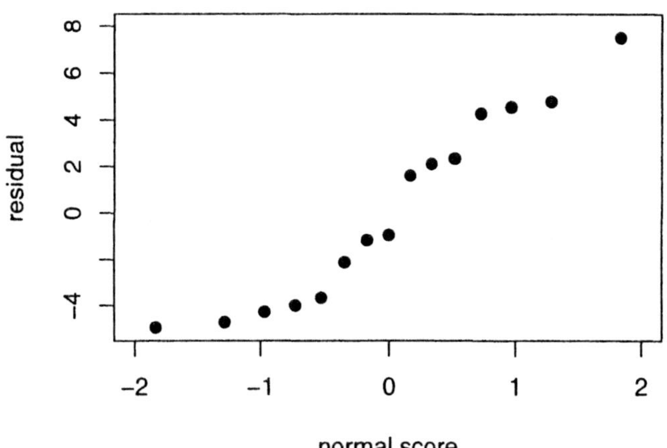

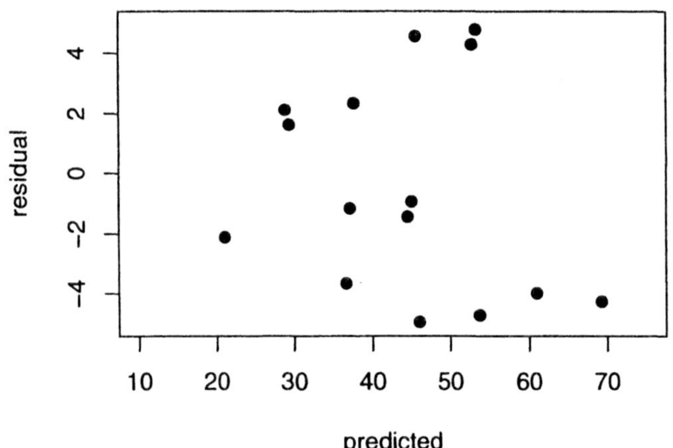

(b) The regression equation is
hardness = 161 + 33.0 Cu - 0.0855 temp

Predictor	Coef	SE Coef	T	P
Constant	161.34	11.43	14.11	0.000
Cu	32.97	16.75	1.97	0.081
temp	-0.085500	0.009788	-8.74	0.000

S = 3.791   R-Sq = 89.9%   R-Sq(adj) = 87.7%

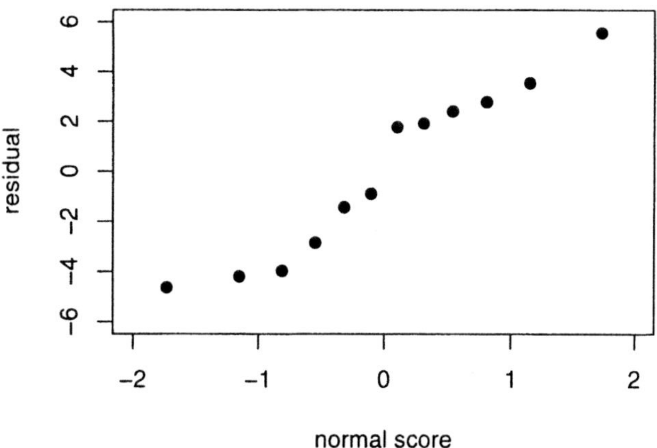

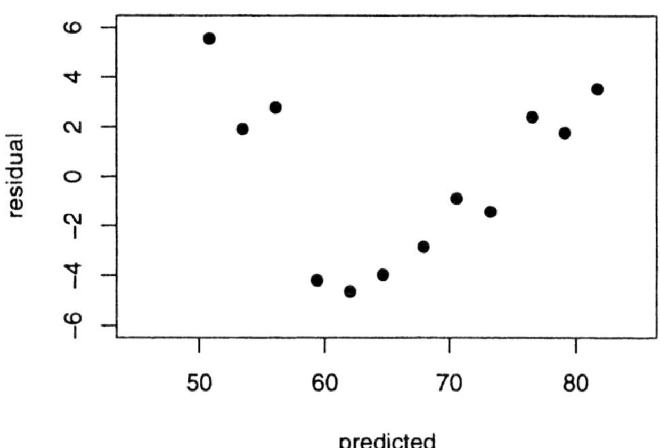

11.45 A plot of the residual versus the time order is given in Figure 11.3. There is a strong increasing pattern indicating a dependence over time. This violates the assumption of independent errors.

11.47 Ordering beer will not increase the number of weddings. There is a lurking variable here, the population size of cities. Large cities will likely have large values for both variables and small cities are likely to have small values of both variables.

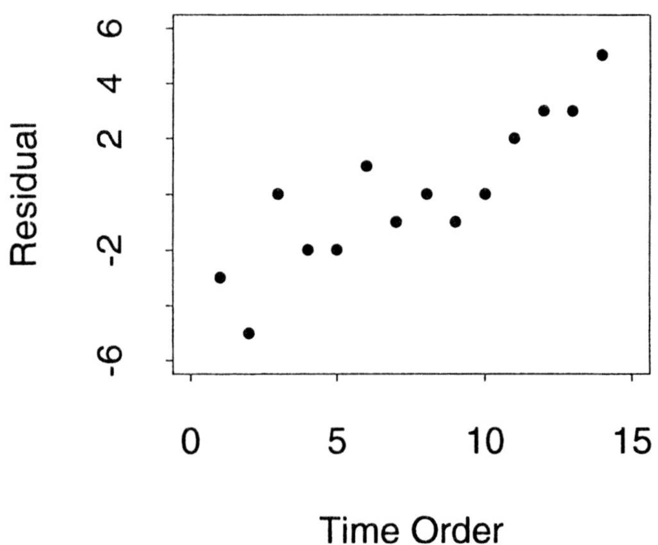

Figure 11.3: Residual Vs. Time Order Plot. Exercise 11.45.

11.49 (a) We first calculate

$$\sum x_i^2 = 532,000, \quad \sum x_i = 2,000, \quad \sum y_i^2 = 9.1097$$

$$\sum y_i = 8.35, \quad \sum x_i y_i = 2175.4$$

and

$$\begin{aligned} S_{xx} &= 532000 - (2000)^2/10 = 132,000 \\ S_{yy} &= 9.1097 - (8.35)^2/10 = 2.13745 \\ S_{xy} &= 2175.4 - (8.35)(2000)/10 = 505.4 \end{aligned}$$

Thus,

$$r \;=\; \frac{505.4}{\sqrt{(132000)(2.13745)}} \;=\; 0.9515.$$

(b) The statistic for testing $\rho = 0$, when the data are from a bivariate normal population is

$$Z \;=\; \frac{\sqrt{n \;-\; 3}}{2}\;\ln\left(\frac{1 \;+\; r}{1 \;-\; r}\right) \;=\; \frac{\sqrt{7}}{2}\;\ln\left(\frac{1 \;+\; .9515}{1 \;-\; .9515}\right) \;=\; 4.89$$

The critical value is 1.96 for a two-sided test with $\alpha = 0.05$, so the null hypothesis that $\rho = 0$ is rejected.

11.51 The statistic for testing the hypothesis that $\rho = 0$,when the data are from a bivariate normal population is

$$Z \;=\; \frac{\sqrt{n-3}}{2}\;\ln\left(\frac{1+r}{1-r}\right) \;=\; \frac{\sqrt{21}}{2}\;\ln\left(\frac{1+.809}{1-.809}\right) \;=\; 5.151$$

The critical value for $\alpha = 0.01$ is 2.56 , so the null hypothesis is rejected.

11.53 The required sums are

$$\sum x \;=\; 533\;,\quad \sum x^2 \;=\; 24529\;,\quad \sum y \;=\; 132$$

$$\sum y^2 \;=\; 1526\;,\quad \sum xy \;=\; 6093$$

$$\begin{aligned} S_{xx} &= 24529 \;-\; (533)^2/12 \;=\; 854.9167 \\ S_{yy} &= 1526 \;-\; (132)^2/12 \;=\; 74 \\ S_{xy} &= 6093 \;-\; (132)(533)/12 \;=\; 230 \end{aligned}$$

Thus,

$$r \;=\; \frac{230}{\sqrt{(854.9167)(74)}} \;=\; 0.914$$

$$\mathcal{Z} = \frac{1}{2}\ln\left(\frac{1+r}{1-r}\right) = \frac{1}{2}\ln\left(\frac{1+.914}{1-.914}\right) = 1.554$$

Since $z_{0.025} = 1.96$, the formula

$$\mathcal{Z} - \frac{z_{0.025}}{\sqrt{n-3}} < \mu_{\mathcal{Z}} < \mathcal{Z} + \frac{z_{0.025}}{\sqrt{n-3}}$$

gives a 95 percent interval from 0.9007 to 2.207 for $\mu_{\mathcal{Z}}$. This corresponds to an interval for ρ from 0.7166 to 0.9761.

11.55 We have $r_1 = 0.83$, $r_1^2 = 0.689$, $r_2 = 0.60$, $r_2^2 = 0.360$. Thus, in the first relationship 68.90 percent of the variation is explained by a linear relationship and in the second relationship 36 percent of the variation is explained by a linear relationship. With regard to variance, the linear relationship in the first case is relatively about twice as strong.

11.57 (a) Using the formula in Exercise 11.53, $\mathcal{Z} = 0.9076$. Then, using the formula in Exercise 11.53 again with $z_{0.025} = 1.96$, the interval for $\mathcal{Z}$ is from 0.4176 to 1.397. The corresponding interval for ρ is from 0.395 to 0.885.

(b) Proceeding as in part (a), $\mathcal{Z} = 0.3654$. The interval for $\mathcal{Z}$ is from -0.0524 to 0.7833. The corresponding interval for ρ is from -0.052 to 0.655.

(c) $\mathcal{Z} = 0.6475$. Thus, the interval for $\mathcal{Z}$ is from 0.3253 to 0.9697. The corresponding interval for ρ is from 0.314 to 0.749.

11.59 (a)

$$f(x,y) = \frac{1}{2\pi\sigma\sigma_1}\exp\left[-\left(\frac{[y-(\alpha+\beta x)]^2}{2\sigma^2} + \frac{(x-\mu_1)^2}{2\sigma_1^2}\right)\right]$$

Rearranging the exponent and making a perfect square in x we obtain,

$$f(x,y) = \frac{1}{\sqrt{2\pi}\sqrt{\sigma^2+\sigma_1^2\beta^2}}\exp\left[-\frac{[y-(\alpha+\beta\mu_1)]^2}{2(\sigma^2+\sigma_1^2\beta^2)}\right] \times$$

$$\frac{\sqrt{\sigma^2+\sigma_1^2\beta^2}}{\sqrt{2\pi}\sigma\sigma_1}\exp\left[-\frac{(\sigma^2+\sigma_1^2\beta^2)}{2\sigma^2\sigma_1^2}\left(x+\frac{\sigma_1^2\alpha\beta-\sigma_1^2\beta y-\sigma^2\mu_1}{\sigma^2+\sigma_1^2\beta^2}\right)^2\right]$$

Now, the marginal density of Y is given by

$$f_2(y) = \int_{-\infty}^{\infty} f(x,y)dx$$

But the first term in the integrand, being independent of x, comes out of the integral and the second term, being a normal density, integrates to 1. So,

$$f_2(y) = \frac{1}{\sqrt{2\pi}\sqrt{\sigma^2+\sigma_1^2\beta_1^2}}\exp\left[-\frac{[y-(\alpha+\beta\mu_1)]^2}{2(\sigma^2+\sigma_1^2\beta_1^2)}\right].$$

This is just a normal density with mean $\alpha+\beta\mu_1$ and variance $\sigma^2+\sigma_1^2\beta^2$. Thus,

$$\mu_2 = \alpha+\beta\mu_1 \text{ and } \sigma_2^2 = \sigma^2+\sigma_1^2\beta^2.$$

A far simpler way to show this is to use the formulas

$$E(Y) = E(E(Y|X)) \text{ and } V(Y) = V(E[Y|X]) + E(V[Y|X]).$$

Now $E(Y|X) = \alpha+\beta X$. So,

$$E(Y) = E(E(Y|X)) = \alpha+\beta E(X) = \alpha+\beta\mu_1.$$

Since $V(Y|X)=\sigma^2$, a constant,

$$V(Y) = V(\alpha+\beta X)+E(\sigma^2) = \beta^2V(X)+\sigma^2 = \beta^2\sigma_1^2+\sigma^2.$$

(b) We start with

$$f(x,y) = \frac{1}{2\pi\sigma\sigma_1}\exp\left[-\left(\frac{[y-(\alpha+\beta x)]^2}{2\sigma^2}+\frac{(x-\mu_1)^2}{2\sigma_1^2}\right)\right]$$

Substituting $\alpha = \mu_2 - \beta\mu_1$, $\sigma^2 = \sigma_2^2 - \beta^2\sigma_1^2$ in the formula, and observing $\rho^2 = 1 - \sigma^2/\sigma_2^2$, we obtain,

$$f(x,y) = \frac{1}{2\pi\sigma_1\sigma_2\sqrt{1-\rho^2}} \times$$

$$\exp\left[-\left(\left(\frac{x-\mu_1}{\sigma_1}\right)^2 - 2\rho\left(\frac{x-\mu_1}{\sigma_1}\right)\left(\frac{y-\mu_2}{\sigma_2}\right) + \left(\frac{y-\mu_2}{\sigma_2}\right)^2\right)/2(1-\rho^2)\right]$$

11.61 The table gives y, $\hat{y}$ and the residuals $y - \hat{y}$

y	$\hat{y}$	$y - \hat{y}$	y	$\hat{y}$	$y - \hat{y}$
38	47.24	−9.24	31	29.69	1.31
40	55.06	−15.06	35	37.51	−2.51
85	62.89	22.11	42	45.34	−3.34
59	70.71	−11.71	59	53.16	5.84
40	38.46	1.54	18	20.91	−2.91
60	46.29	13.71	34	28.74	5.26
68	54.11	13.89	29	36.56	−7.56
53	61.94	−8.94	42	44.39	−2.39

(a) $\sum(y - \bar{y})^2 = 4318.44$

(b) $\sum(y - \hat{y})^2 = 1553.81$

(c) $r = \sqrt{1 - (1553.81)/(4318.44)} = 0.80.$

11.63 (a) Written as ranks, the data in this example are

x	y	d_i^2	x	y	d_i^2
2	1	1	10	9	1
1	4	9	9	5	16
3	3	0	5	8	9
7	10	9	6	6	0
4	2	4	8	7	1

$$r_S = 1 - \frac{6\sum d_i^2}{n(n^2 - 1)} = 1 - \frac{6(50)}{10(100 - 1)} = 0.697.$$

This is a little smaller than the 0.73 obtained using the sample correlation.

(b) Written as ranks the data in this example are

x	y	d_i^2	x	y	d_i^2
16	10	36	10.5	15	20.25
3	1	4	13	11.5	2.25
17	21	16	10.5	2	72.25
6	13	49	21.5	19	6.25
6	3	9	1.5	6.5	25
21.5	23	2.25	14	21	49
8.5	8	0.25	15	18	9
4	6.5	6.25	6	4.5	2.25
23.5	17	42.25	19.5	15	20.25
8.5	11.5	9	1.5	4.5	9
19.5	24	20.25	23.5	21	6.25
12	9	9	18	15	9

Thus $\sum d_i^2 = 434$ and

$$r_S = 1 - \frac{6 \sum d_i^2}{n(n^2 - 1)} = 1 - \frac{6(434)}{24(24^2 - 1)} = 0.811.$$

This is essentially the same as the .809 obtained from the sample correlation.

(c) Written as ranks, the data in this example are

x	y	d_i^2	x	y	d_i^2
1	1	0	6	5	1
2	3	1	7	8	1
3	2	1	8	9	1
4	6	4	9	7	4
5	4	1	10	10	0

Thus $\sum d_i^2 = 14$ and

$$r_S = 1 - \frac{6 \sum d_i^2}{n(n^2 - 1)} = 1 - \frac{6(14)}{10(100 - 1)} = 0.915.$$

This is also close to, though less than $r = 0.9629$.

11.65 When $\alpha = 0.05$, the two-sided critical value is 1.96.

$$\sigma_{r_S} = \frac{1}{\sqrt{17}} = 0.2425.$$

$$1.96\ \sigma_{r_S} = 0.475.$$

Since $r_S = 0.39 < 0.475$, r_S is not significantly different from 0.

11.67
```
Pearson correlation of resistance and time = 0.809
P-Value = 0.000
```

11.69 The variance estimate is

$$s_e^2 = \frac{10(4600) - (200)^2}{(5-2)(10)} = 200$$

The 95 percent confidence interval for α when $t_{.025} = 3.182$ for 3 degrees of freedom is given by

$$-10 \pm 3.182\sqrt{200}\sqrt{\frac{10+45}{5(10)}}$$

Thus, the interval is from -37.2 to 57.2.

11.71 (a) The prediction for the mean repair time at $x = 4.5$ is

$$\hat{y} = a + b(4.5) = 2 + 16(4.5) = 74$$

and the 95 percent confidence interval is

$$74 \pm 3.182\sqrt{680}\sqrt{\frac{1}{5} + \frac{(4.5-3)^2}{10}}.$$

So the interval is from 19.9 hours to 128.1 hours.

(b) The 95 percent limits of prediction for a single engine that is run $x = 4.5$

thousand hours is

$$74 \pm 3.182\sqrt{680}\sqrt{1+\frac{1}{5}+\frac{(4.5-3)^2}{10}}.$$

So the interval is from 0 hours to 173.1 hours (the lower limit of -25.1 does not make any sense, so we take 0 as the lower limit).

11.73 The variance estimate is $s_e = .4154$ by (c) of 11.68. With $t_{.025} = 2.306$, a 95 percent confidence interval for α is given by

$$.619 \pm 2.306(.4154)\sqrt{\frac{1}{10}+\frac{3.5^2}{124.375}} = .619 \pm .427$$

Thus, the interval for α is from .192 to 1.046.

11.75 The ideal gas law is $pV^{\gamma} = C$. Taking logs of both sides gives

$$\ln p + \gamma \ln V = C.$$

Let $y_i = \ln p_i$ and $x_i = -\ln V_i$. Then, $\bar{x} = -2.815$, $\bar{y} = 4.530$, $S_{xx} = 3.3068$, $S_{xy} = 4.9573$. So,

$$\begin{aligned} \gamma &= \frac{S_{xy}}{S_{xx}} = 1.499 \\ C &= \bar{y} - b\bar{x} = 8.760 \end{aligned}$$

11.77 For the equation $I = 1 - e^{-\epsilon/\tau}$,

$$\ln(1-I) = -\frac{1}{\tau}t.$$

Let $y_i = \ln(1-I_i)$. Notice that this equation does not have an intercept, so use

$$b = \frac{\sum_{i=1}^{n} y_i t_i}{\sum_{i=1}^{n} t_i^2}.$$

$$\sum_{i=1}^{n} y_i t_i = -2.1793, \quad \sum_{i=1}^{n} t_i^2 = 2.04.$$

So

$$b = \frac{-2.1793}{2.04} = -1.068 \quad \text{and} \quad \tau = -\frac{1}{b} = 0.936.$$

11.79 (a) From Exercise 11.9, $S_{xx} = 10$, $S_{xy} = 20$, $S_{yy} = 46$ and $\hat{y} = 0 + 2x$. Also $s_e^2 = 2.00$. The prediction for the mean CPU time at $x = 3.0$ is $\hat{y} = 0 + 2(3.0) = 6$. The 95 percent confidence interval is given by

$$6 \pm 3.182\sqrt{2.00}\sqrt{\frac{1}{5} + \frac{(3-3)^2}{10}}.$$

So the interval is from 3.99 to 8.01 hours.

(b) The 95 percent limits of prediction for a single future day is

$$6 \pm 3.182\sqrt{2.00}\sqrt{1 + \frac{1}{5} + \frac{(3-3)^2}{10}}.$$

So the interval is from 1.07 to 10.93 hours.

11.81 $r_1 = 0.41$, $r_1^2 = 0.1681$, $r_2 = 0.29$, $r_2^2 = 0.0841$. Thus, in the first relationship 16.81 percent of the variation is explained by a linear relationship and in the second relationship 8.41 percent of the variation is explained by a linear relationship. With regard to variance, the linear relationship in the first case is relatively about twice as strong.

11.83 (a) Since the transformation $\mathcal{Z} = (1/2)\ln(1 + r/1 - r) = 1.045$ and $z_{0.025} = 1.96$, the 95% confidence interval for $\mathcal{Z}$ is $\mathcal{Z} \pm 1.96/\sqrt{n-3}$ or from 0.4796 to 1.6112. The corresponding interval for ρ is from 0.446 to 0.923 .

(b) Proceeding as in part (a), $\mathcal{Z} = -0.725$. The interval for $\mathcal{Z}$ is from -1.089 to -0.361. The corresponding interval for ρ is from -0.797 to -0.346.

(c) $\mathcal{Z} = 0.1717$. Thus, the interval for $\mathcal{Z}$ is from -0.1748 to 0.5181. The corresponding interval for ρ is from -0.173 to 0.476.

11.85 (a) We have $n = 26$

$$\begin{aligned} \textstyle\sum_{i=1}^n x_i &= 102, & \textstyle\sum_{i=1}^n x_i^2 &= 428, \\ \textstyle\sum_{i=1}^n y_i &= 47.057, & \textstyle\sum_{i=1}^n x_i y_i &= 197.99. \end{aligned}$$

so

$$\begin{aligned} S_{xx} &= \sum_{i=1}^n x_i^2 - (\sum_{i=1}^n x_i)^2/n = 428 - (102)^2/26 = 27.8462 \\ S_{xy} &= \sum_{i=1}^n x_i y_i - (\sum_{i=1}^n x_i)(\sum_{i=1}^n y_i)/n = 197.99 - 102(47.057)/26 = 13.3817 \end{aligned}$$

Consequently,

$$\begin{aligned} b &= \frac{S_{xy}}{S_{xx}} = \frac{13.382}{27.846)} = .048 \\ a &= \overline{y} - b\overline{x} = \frac{47.057}{26} - \frac{13.3817}{27.8462}\frac{102}{26} = -.075 \end{aligned}$$

So, the least squares line is

$$\hat{y} = -0.075 + 0.48x.$$

(b) In addition to the sums $S_{xx} = 27.846$ and $S_{xy} = 13.382$ from Part (a), we calculate

$$S_{yy} = \sum_{i=1}^n y_i^2 - (\sum_{i=1}^n y_i)^2/n = 91.794 - (47.057)^2/26 = 6.626,$$

Thus,

$$\begin{aligned} s_e^2 &= \frac{S_{xx}S_{yy} - (S_{xy})^2}{(n-2)S_{xx}} \\ &= \frac{(27.846)(6.626) - (13.382)^2}{(24)(27.846)} = .0081245601 \end{aligned}$$

The 95 percent confidence interval is given by

$$b \pm t_{\frac{\alpha}{2}} s_e \sqrt{\frac{1}{S_{xx}}}$$

or, since $t_{.025}$ for 24 degrees of freedom is 2.064,

$$0.48 \pm 2.064\sqrt{.0081245601}\sqrt{\frac{1}{27.846}}$$

The 95% confidence interval is from .035 to .515 thousandths of an inch per thousand pounds.

(c) We have to test the null hypothesis $H_0 : \alpha = 0$ against the alternative $H_1 : \alpha \neq 0$ at 0.05 level of significance. The t statistic is given by

$$t = \frac{a - \alpha}{s_e}\sqrt{\frac{nS_{xx}}{S_{xx} + n\bar{x}^2}}.$$

In our case, $S_{xx} = 27.846$ and $a = -0.07511$. So,

$$t = \frac{(-0.07511 - 0)}{\sqrt{.0081245601}}\sqrt{\frac{26(27.846)}{27.846 + 26(3.923)^2}} = 1.4096.$$

Since $t_{.025}$ is 2.064 for 24 degrees of freedom, we cannot reject the null hypothesis $H_0 : \alpha = 0$.

(d) Figure 11.4 shows the residuals plotted against the fitted values.

The residual plot does not reveal any pattern and hence we can conclude that the assumptions of the model are not violated.

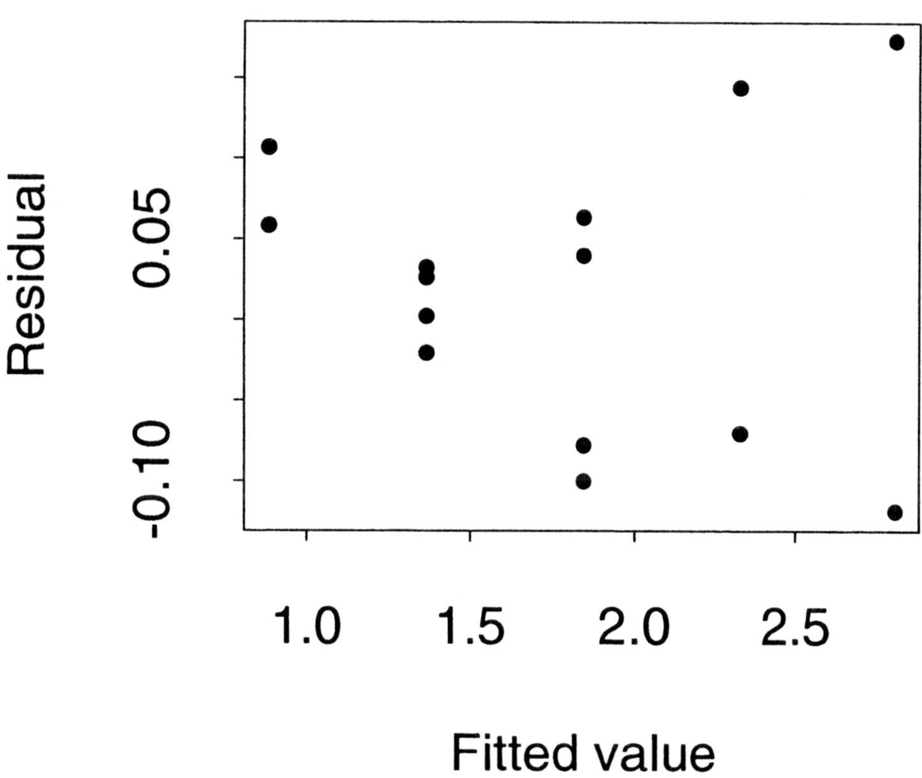

Figure 11.4: Residual Plot for Exercise 11.85

Chapter 12

ANALYSIS OF VARIANCE

12.1 (a) If the experiment is performed in soft water, the results may only be valid in soft water. Other kinds of water should also be used.

(b) With 15 results for detergent A and only 5 for detergent B, the variability for detergent A is known with much more precision than that for detergent B. Equal sample sizes should be used.

(c) The results may only apply to very hot water and very short wash times, and not to the circumstances of normal use with low temperature and longer washing times.

(d) There may be a time effect for the measurement process determining "whiteness". for example, the instrument may require recalibration after a few readings. The test results may be biased in this case.

12.3 (a) The grand mean is $\bar{y}_{\cdot} = 78$. The deviations from the grand mean are:

Method A	Method B	Method C
−5	13	−6
−1	3	−1
−11	9	−2
−17	7	1

The sum of squares of the deviations is

$$SST = \sum_{i=1}^{k} \sum_{j=1}^{n_i} (y_{ij} - \bar{y}_{\cdot})^2 = 546.$$

The means for the three samples are

Method A	Method B	Method C
72	86	76

The deviations of each sample from its own mean are

Method A	Method B	Method C
1	5	− 4
5	−5	1
−5	1	0
−1	− 1	3

The sum of squares of these deviations is

$$SSE = \sum_{i=1}^{k} \sum_{j=1}^{n_i} (y_{ij} - \bar{y}_j)^2 = 130.$$

The deviation of the individual sample means from the grand mean are

$$-6 \quad 8 \quad -2$$

Thus,

$$SS(Tr) = \sum_{i=1}^{k} n_i(\bar{y}_i - \bar{y}_.)^2 = 4(36) + 4(64) + 4(4) = 416.$$

(b) The total sample size $N = 4 + 4 + 4 = 12$, and the totals of each sample are

Method A	Method B	Method C
288	344	304

Thus, $T_. = 288 + 344 + 304 = 936$,

$$C = \frac{936^2}{121} = 73,008 \quad \text{and} \quad \sum_{i=1}^{k} \frac{T_i^2}{n_i} = 73,424$$

The sum of squares of all of the observations is 73,554. Thus,

$$SST = \sum_{i=1}^{k} \sum_{j=1}^{n_i} y_{ij}^2 - C = 73,554 - 73,008 = 546.$$

$$SS(Tr) = \frac{\sum_{i=1}^{k} T_i^2}{n_i} - C = 73,424 - 73,008 = 416.$$

and

$$SSE = SST - SS(Tr) = 130.$$

These numbers agree with part (a).

12.5 The null hypothesis is that the mean number of mistakes is the same for the four technicians. The alternative is that the means are not the same. The analysis-of-variance table is

Source of variation	Degrees of freedom	Sum of squares	Mean square	F
Technicians	3	12.95	4.3167	0.68
Error	16	101.60	6.3500	
Total	19	114.55		

Since the critical value at the 0.01 level for an F distribution with 3 and 16 degrees of freedom is 5.29, we cannot reject the null hypothesis.

12.7 (a) We first find $k = 4$, $\bar{y} = 7$, $\bar{y}_1 = 5$, $\bar{y}_2 = 12$, $\bar{y}_3 = 9$, and $\bar{y}_4 = 3$. Thus,

$$\begin{array}{c} \text{Obs.} \\ y_{ij} \\ \left[\begin{array}{ccccc} 6 & 4 & 5 & & \\ 13 & 10 & 13 & 12 & \\ 7 & 9 & 11 & & \\ 3 & 6 & 1 & 4 & 1 \end{array}\right] \end{array} = \begin{array}{c} \text{Grand mean} \\ \bar{y} \\ \left[\begin{array}{ccccc} 7 & 7 & 7 & & \\ 7 & 7 & 7 & 7 & \\ 7 & 7 & 7 & & \\ 7 & 7 & 7 & 7 & 7 \end{array}\right] \end{array}$$

$$+ \begin{array}{c} \text{Tr. effect} \\ \bar{y}_i - \bar{y} \\ \left[\begin{array}{ccccc} -2 & -2 & -2 & & \\ 5 & 5 & 5 & 5 & \\ 2 & 2 & 2 & & \\ -4 & -4 & -4 & -4 & -4 \end{array}\right] \end{array} + \begin{array}{c} \text{Error} \\ y_{ij} - \bar{y}_i \\ \left[\begin{array}{ccccc} 1 & -1 & 0 & & \\ 1 & -2 & 1 & 0 & \\ -2 & 0 & 2 & & \\ 0 & 3 & -2 & 1 & -2 \end{array}\right] \end{array}$$

$$\begin{aligned} SS(Tr) &= 3(-2)^2 + 4(5)^2 + 3(2)^2 + 5(-4)^2 = 204, \quad df = 3. \\ SSE &= 1^2 + (-1)^2 + \cdots + 1^2 + (-2)^2 = 34, \quad df = 15 - 4 = 11. \\ SST &= 6^2 + 4^2 + \cdots + 4^2 + 1^2 - 15(7)^2 = 238, \quad df = 14 \\ &= SS(Tr) + SSE = 238. \ (\textit{check}) \end{aligned}$$

(b) The null hypothesis is that the the four treatment population means are the same. The alternative is that they are not the same. The analysis-of-variance table is

Source of variation	Degrees of freedom	Sum of squares	Mean square	F
Treatments	3	204	68	22.00
Error	11	34	3.89	
Total	14	238		

Since the critical value at the 0.05 level for an F distribution with 3 and 11 degrees of freedom is 3.59, we can reject the null hypothesis. The treatment means are not the same.

12.9 The null hypothesis is that the true means of the samples of reactions times are the same. The alternative is that the true means are not the same. The analysis-of-variance table is

Source of variation	Degrees of freedom	Sum of squares	Mean square	F
Arrangements	2	81.429	40.7145	11.310
Error	25	90.000	3.600	
Total	27	171.429		

Since the critical value at the 0.01 level for an F distribution with 2 and 25 degrees of freedom is 5.57, we can reject the null hypothesis. The mean reaction times are not the same for the three arrangements.

12.11 The null hypothesis is that the true means of the samples of the six samples are the same. The alternative is that the true means are not the same. The analysis-of-variance table is

Source of variation	Degrees of freedom	Sum of squares	Mean square	F
Samples	5	1569.667	313.9334	15.697
Error	6	120.000	20.0000	
Total	11	1689.667		

Since the critical value at the 0.05 level for an F distribution with 5 and 6 degrees of freedom is 4.39, we reject the null hypothesis. These samples differ in mean compressive strength.

12.13 By assumption

$$P\left(|(\bar{Y}_i - \bar{Y}_j) - (\mu_i - \mu_j) > |E|\right) \geq \alpha\,, \qquad 1 \leq i \neq j \leq 3$$

Hence, for the equal tail test,

$$P\left(|Z| > E/\sigma\sqrt{\frac{1}{n_i} + \frac{1}{n_j}}\right) \geq \alpha$$

which implies

$$E/\sigma\sqrt{\frac{1}{n_i} + \frac{1}{n_j}} = z_{\alpha/2}$$

or

$$\frac{1}{n_i} + \frac{1}{n_j} = \left(\frac{E}{z_{\alpha/2}\sigma}\right)^2 = K \quad \text{where} \quad 1 \leq i \neq j \leq 3$$

That is, we require that

$$\frac{1}{n_1} = \frac{1}{n_2} = \frac{1}{n_3} = \frac{K}{2}$$

or

$$n_1 = n_2 = n_3 = \frac{2}{K} = 2\left(\frac{z_{\alpha/2}\sigma}{E}\right)^2 = 2\left(\frac{(1.96)(0.012)}{0.01}\right)^2 = 11.06$$

Therefore we take $n_1 = n_2 = n_3 = 12$.

12.15 By Theorem 12.1, using the notation $N = \sum_{i=1}^{k} n_i$,

$$\begin{aligned}
SST &= \sum_{i=1}^{k}\sum_{j=1}^{n_i}(y_{ij} - \bar{y}_{\cdot})^2 = \sum_{i=1}^{k}\sum_{j=1}^{n_i}(y_{ij}^2 - 2y_{ij}\bar{y}_{\cdot} + \bar{y}_{\cdot}^2) \\
&= \sum_{i=1}^{k}\sum_{j=1}^{n_i} y_{ij}^2 - 2\bar{y}_{\cdot}\sum_{i=1}^{k}\sum_{j=1}^{n_i} y_{ij} + N\bar{y}_{\cdot}^2 \\
&= \sum_{i=1}^{k}\sum_{j=1}^{n_i} y_{ij}^2 - N\bar{y}_{\cdot}^2 = \sum_{i=1}^{k}\sum_{j=1}^{n_i} y_{ij}^2 - C
\end{aligned}$$

since $N\bar{y}_.^2 = T_.^2/N$.

Similarly,

$$\begin{aligned} SS(Tr) &= \sum_{i=1}^{k} n_i(\bar{y}_i - \bar{y}_.)^2 = \sum_{i=1}^{k} n_i\bar{y}_i^2 - 2\bar{y}_. \sum_{i=1}^{k} n_i\bar{y}_i + N\bar{y}_.^2 \\ &= \sum_{i=1}^{k} n_i \left(\frac{T_i}{n_i}\right)^2 - 2\frac{T_.}{N}\sum_{i=1}^{k} n_i \frac{T_i}{n_i} + N\left(\frac{T_.}{N}\right)^2 \\ &= \sum_{i=1}^{k} \frac{T_i^2}{n_i} - 2\frac{T_.}{N}\sum_{i=1}^{k} T_i + \frac{T_.^2}{N} \\ &= \sum_{i=1}^{k} \frac{T_i^2}{n_i} - \frac{T_.^2}{N} = \sum_{i=1}^{k} \frac{T_i^2}{n_i} - C \end{aligned}$$

12.17 (a) The null hypothesis is that the true mean aflatoxin content is the same for both brands. The alternative is that they are different. The analysis-of-variance table is

Source of variation	Degrees of freedom	Sum of squares	Mean square	F
Brands	1	11.734	11.7340	1.047
Error	12	134.515	11.2096	
Total	13	146.249		

Since the critical value at the 0.05 level for an F distribution with 1 and 12 degrees of freedom is 4.75, we cannot reject the null hypothesis. The two brands do not differ in mean aflatoxin content.

(b) The two-sample t statistic for testing the null hypothesis $\delta = 0$ versus the alternative hypothesis $\delta \neq 0$, is given by

$$t = \frac{\bar{y}_1 - \bar{y}_2}{\sqrt{(n_1 - 1)s_1^2 + (n_2 - 1)s_2^2}} \sqrt{\frac{n_1 n_2(n_1 + n_2 - 2)}{n_1 + n_2}}$$

In this case $\bar{y}_1 = 2.2$, $\bar{y}_2 = 4.05$, $s_1^2 = 8.157$ and $s_2^2 = 15.483$. Since $n_1 = 8$

and $n_2 = 6$,

$$t = \frac{2.2 - 4.05}{\sqrt{(8-1)8.157 + (6-1)15.483}} \sqrt{\frac{(8)(6)(8+6-2)}{8+6}} = -1.023$$

Since the .025 value of a t with 12 degrees of freedom is 2.179. we cannot reject the null hypothesis.

(c) Notice that, within roundoff, $(-1.023)^2 = 1.047$ for the statistics and that $(2.179)^2 = 4.75$ for the critical values. The two analyzes are equivalent, since both statistics will reject and accept at the same time.

12.19 (a) There are $a = 3$ treatments the degrees of freedom for Total are $ab - 1 = 11$ so $b = 12/3 = 4$. The Block sum of squares has degrees of freedom $= b - 1 = 4 - 1 = 3$. Next,

$$\overline{y}_{..} = \frac{4\overline{y}_{1.} + 4\overline{y}_{2.} + 4\overline{y}_{3.}}{3 \times 4} = \frac{96}{12} = 8$$

so the sum of squares treatment

$$b \sum_{i=1}^{a} (\overline{y}_{i.} - \overline{y}_{..}) = 4(-2)^2 + 4(-1)^2 + 4(3)^2 = 56$$

The treatment sum of squares has df $= a - 1 = 3 - 1 = 2$.

Finally, by subtraction, the error sum of squares

$$SSE = SST - SS(Tr) - SS(Bl) = 220 - 56 - 132 = 32$$

with $(a-1)(b-1) = 2 \cdot 3 = 6$ degrees of freedom

The completed analysis-of-variance table is

Source of variation	Degrees of freedom	Sum of squares	Mean square	F
Treatment	2	56	28	5.250
Blocks	3	132	44	8.251
Error	= 6	32	5.333	
Total	11	220		

(b) Since the critical value at the 0.05 level for an F distribution with 2 and 6 degrees of freedom is 5.14, we reject the null hypothesis of no differences in the treatment means. Further, since the critical value at the 0.05 level for an F distribution with 3 and 6 degrees of freedom is 4.76, we reject the null hypothesis of no differences in blocks.

12.21 The null hypothesis is that the true means for the technicians are the same. The alternative hypothesis is that they are not the same. A second null hypothesis is that there is no block(day) effect.

The analysis-of-variance table is

Source of variation	Degrees of freedom	Sum of squares	Mean square	F
Technicians	3	12.95	4.32	8.642
Days	4	20.80	5.20	0.773
Error	12	80.8	6.73	
Total	19	114.55		

Since the critical value at the 0.01 level for an F distribution with 3 and 12 degrees of freedom is 5.95, we cannot reject the null hypothesis that there is no difference between the means for the technicians. Since the critical value at the 0.01 level for an F distribution with 4 and 12 degrees of freedom is 5.41, we cannot reject the null hypothesis of no block effect.

12.23 (a) We first find $a = 4$, $\bar{y}_{..} = 10$, $\bar{y}_{1.} = 8$, $\bar{y}_{2.} = 13$, $\bar{y}_{3.} = 10$ and $\bar{y}_{4.} = 9$. And $b = 5$, $\bar{y}_{.1} = 13$, $\bar{y}_{.2} = 8$, $\bar{y}_{.3} = 12$, $\bar{y}_{.4} = 6$ and $\bar{y}_{.5} = 11$. Thus,

$$\underset{y_{ij}}{\text{Obs.}}\quad \begin{bmatrix} 14 & 6 & 11 & 0 & 9 \\ 14 & 10 & 16 & 9 & 16 \\ 12 & 7 & 10 & 9 & 12 \\ 12 & 9 & 11 & 6 & 7 \end{bmatrix} = \underset{\bar{y}_{..}}{\text{Grand mean}}\quad \begin{bmatrix} 10 & 10 & 10 & 10 & 10 \\ 10 & 10 & 10 & 10 & 10 \\ 10 & 10 & 10 & 10 & 10 \\ 10 & 10 & 10 & 10 & 10 \end{bmatrix}$$

$$\underset{\bar{y}_{i.} - \bar{y}_{..}}{\text{Tr. effect}}\quad + \begin{bmatrix} -2 & -2 & -2 & -2 & -2 \\ 3 & 3 & 3 & 3 & 3 \\ 0 & 0 & 0 & 0 & 0 \\ -1 & -1 & -1 & -1 & -1 \end{bmatrix} \quad \underset{\bar{y}_{.j} - \bar{y}_{..}}{\text{Bl. effect}}\quad + \begin{bmatrix} 3 & -2 & 2 & -4 & 1 \\ 3 & -2 & 2 & -4 & 1 \\ 3 & -2 & 2 & -4 & 1 \\ 3 & -2 & 2 & -4 & 1 \end{bmatrix}$$

$$\underset{y_{ij} - \bar{y}_{i.} - \bar{y}_{.j} + \bar{y}_{..}}{\text{Error}}\quad + \begin{bmatrix} 3 & 0 & 1 & -4 & 0 \\ -2 & -1 & 1 & 0 & 2 \\ -1 & -1 & -2 & 3 & 1 \\ 0 & 2 & 0 & 1 & -3 \end{bmatrix}$$

(b) The sums of squares and degrees of freedoms are

$$\begin{aligned} SS(Tr) &= 5(-2)^2 + 5(3)^2 + 5(0)^2 + 5(-1)^2 = 70, \quad df = 3. \\ SS(Bl) &= 4(3)^2 + 4(-2)^2 + 4(2)^2 + 4(-4)^2 + 4(1) = 136, \quad df = 4. \\ SSE &= 3^2 + 0^2 + 1^2 + \cdots + 0^2 + 1^2 + (-3)^2 = 66, \\ & \qquad df = (4-1)(5-1) = 12. \end{aligned}$$

$$
\begin{aligned}
SST &= 14^2 + 6^2 + 11^2 + \cdots + 11^2 + 6^2 + 7^2 - 20(10)^2 = 272, \\
& \qquad df = 19 \\
&= SS(Tr) + SS(Bl) + SSE = 272. \quad (check)
\end{aligned}
$$

(c) The null hypothesis is that the the four treatment population means are the same. The alternative is that they are not the same. The analysis-of-variance table is

Source of variation	Degrees of freedom	Sum of squares	Mean square	F
Treatments	3	70	23.33	4.24
Blocks	4	136	34.00	6.18
Error	12	66	5.50	
Total	19	272		

Since the critical value at the 0.05 level for an F distribution with 3 and 12 degrees of freedom is 3.49, we can reject the null hypothesis. The treatment means are not the same. Since $F_{.05}$ with 4 and 12 degrees of freedom is 3.26, the block effect of the experiment is also apparent.

12.25 The null hypothesis is that the true means for the detergents are the same. The alternative hypothesis is that they are not the same.

The one-way analysis-of-variance table is

Source of variation	Degrees of freedom	Sum of squares	Mean square	F
Detergents	3	110.92	36.97	1.95
Error	8	154.00	19.25	
Total	11	264.92		

Since the critical value at the 0.05 level for an F distribution with 3 and 8 degrees of freedom is 4.07, we cannot reject the null hypothesis of no detergent effect. In the two-way analysis, we could reject the null hypothesis at the $\alpha = 0.01$ level. This shows how important it is to be sure that there is no systematic variation in the error sum of squares.

12.27 To use the sums of squares formula, we need to compute the marginal and total sums

$$T_{..1} = 438, \quad T_{..2} = 483$$
$$T_{.1.} = 223, \quad T_{.2.} = 236, \quad T_{.3.} = 242, \quad T_{.4.} = 220$$
$$T_{1..} = 308, \quad T_{2..} = 304, \quad T_{3..} = 309, \quad T_{...} = 921$$

Thus, $C = (921)^2/(3 \cdot 4 \cdot 2) = 35,343.375$. Further,

$$\sum_{i=1}^{a} T_{i..}^2 = 308^2 + 304^2 + 309^2 = 282,761$$

$$\sum_{j=1}^{b} T_{.j.}^2 = 223^2 + 236^2 + 242^2 + 220^2 = 212,389$$

$$\sum_{k=1}^{r} T_{..k}^2 = 438^2 + 483^2 = 425,133$$

$$\sum_{i=1}^{a}\sum_{j=1}^{b}\sum_{k=1}^{r} y_{ijk}^2 = 35,715$$

Thus,

$$SST = 35,715 - 35,343.375 = 371.625$$

$$SS(Tr) = \frac{282,761}{4 \cdot 2} - 35,343.375 = 1.750$$

$$SS(Bl) = \frac{212,389}{3 \cdot 2} - 35,343.375 = 54.792$$

$$SS(Reps) = \frac{425,133}{4 \cdot 3} - 35,343.375 = 84.375$$

$$SSE = 371.62 - 1.745 - 54.79 - 84.37 = 230.715$$

The analysis of variance is

Source of variation	Degrees of freedom	Sum of squares	Mean square	F
Machines	2	1.750	0.875	0.064
Workers	3	54.792	18.264	1.364
Reps	1	84.375	84.375	6.217
Error	17	230.708	13.571	
Total	23	371.625		

Since the critical value at the 0.05 level for an F distribution with 2 and 17 degrees of freedom is 3.59, we cannot reject the null hypothesis of no treatment (machine) effect.

Since the critical value at the 0.05 level for an F distribution with 3 and 17 degrees of freedom is 3.20, we cannot reject the null hypothesis of no block(worker) effect.

Since the critical value at the 0.05 level for an F distribution with 1 and 17 degrees of freedom is 4.45 , we reject the null hypothesis of no replication effect.

12.29 We are given that

$$\mu_{ij} = \mu + \alpha_i + \beta_j, \qquad \frac{1}{b}\sum_{j=1}^{b}\mu_{ij} = \mu + \alpha_i, \qquad \frac{1}{ab}\sum_{i=1}^{a}\sum_{j=1}^{b}\mu_{ij} = \mu$$

Summing the first relation over j and dividing by b gives

$$\frac{1}{b}\sum_{j=1}^{b}\mu_{ij} = \mu + \alpha_i + \frac{1}{b}\sum_{j=1}^{b}\beta_j$$

Comparing this with the second given relation, we conclude that

$$\frac{1}{b}\sum_{j=1}^{b}\beta_j = 0 \ \ or \ \ \sum_{j=1}^{b}\beta_j = 0$$

Next, summing the first given relation over i and j and dividing by ab,

$$\frac{1}{ab}\sum_{i=1}^{a}\sum_{j=1}^{b}\mu_{ij} = \mu + \frac{1}{a}\sum_{i=1}^{a}\alpha_i + \frac{1}{b}\sum_{j=1}^{b}\beta_j$$

Comparing this with the third given relation, and using $\sum_{j=1}^{b}\beta_j = 0$, we conclude that $\sum_{k=1}^{a}\alpha_i = 0$,

12.31 From Exercise 12.21, the (sorted) means for the four treatments are

Treatment 1	Treatment 4	Treatment 3	Treatment 2
8	9	10	13

and the $MSE = 5.5$ with 5 degrees of freedom. From Table 12(a), after multiplying each r_p by $s_{\bar{y}} = \sqrt{MSE/n} = \sqrt{5.5/5} = 1.049$ to get R_p, we obtain

p	2	3	4
r_p	3.08	3.23	3.31
R_p	3.23	3.39	3.47

For $\alpha = .05$, none of the ranges between two adjacent means are significant. Only $\bar{y}_2 - \bar{y}_4 = 4$ is significant among the ranges of three adjacent means. The range of four means $\bar{y}_2 - \bar{y}_1 = 5$ is significant among the ranges of three adjacent means. The range of four means $\bar{y}_2 - \bar{y}_1 = 5 > R_4$ is also significant. We conclude

Tr 1	Tr 4	Tr 3	Tr 2
8	9	10	13

12.33 The (sorted) means for the three methods are

A	C	B
72	76	86

and the $MSE = 130/9 = 14.44$ with 9 degrees of freedom. From Table 12(a), after multiplying each r_p by $s_{\bar{y}} = \sqrt{MSE/n} = \sqrt{14.44/4} = 1.9$ to get R_p, we obtain

p	2	3
r_p	3.20	3.34
R_p	6.08	6.35

For $\alpha = .05$, we conclude

A	C	B
72	76	86

12.35 (a) The (sorted) sample means for the five threads are

thread 1	thread 5	thread 3	thread 4	thread 2
20.675	20.900	23.525	23.700	25.650

and, from Exercise 12.18, the $MSE = 2.110$ with 12 degrees of freedom. From Table 12(b) with 12 degrees of freedom , after multiplying each r_p by $s_{\bar{y}} = \sqrt{MSE/n} = \sqrt{2.110/4} = 0.726$ to get R_p, we obtain

p	2	3	4	5
r_p	4.32	4.50	4.62	4.71
R_p	3.14	3.27	3.35	3.42

We conclude that, at the .01 level of significance, the means of detergents B and C are significantly different but that the means of detergents A and B are not different.

thread 1	thread 5	thread 3	thread 4	thread 2
20.675	20.900	23.525	23.700	25.650

(b) With $k = 5$, $\alpha = .10$ and $\alpha/k(k-1) = .10/20 = .005$, we have $t_{.005} = 3.055$ with 12 degrees of freedom. Using Bonferroni's procedure, we can construct 10 confidence intervals for the differences of mean breaking strength of the five threads and are 90% confidence that all 10 intervals hold simultaneously. For example, the confidence interval for $\mu_1 - \mu_2$ is

$$\mu_1 - \mu_2 \ : \qquad (\bar{y}_{1.} - \bar{y}_{2.}) \pm t_{.005}\sqrt{MSE(\frac{1}{4} + \frac{1}{4})}$$

$$= (20.675 - 25.650) \pm 3.055\sqrt{2.11(\frac{1}{4} + \frac{1}{4})}$$
$$= -4.975 \pm 3.138$$

Similarly, with the same 3.138 applying for all pairs of mean differences, we obtain

$$\begin{array}{lll}
\mu_1 - \mu_3: & (20.675 - 23.525) \pm 3.138 & = -2.850 \pm 3.138 \\
\mu_1 - \mu_4: & (20.675 - 23.700) \pm 3.138 & = -3.025 \pm 3.138 \\
\mu_1 - \mu_5: & (20.675 - 20.900) \pm 3.138 & = -0.225 \pm 3.138 \\
\mu_2 - \mu_3: & (25.650 - 23.525) \pm 3.138 & = 2.125 \pm 3.138 \\
\mu_2 - \mu_4: & (25.650 - 23.700) \pm 3.138 & = 1.950 \pm 3.138 \\
\mu_2 - \mu_5: & (25.650 - 20.900) \pm 3.138 & = 4.750 \pm 3.138 \\
\mu_3 - \mu_4: & (23.525 - 23.700) \pm 3.138 & = -0.175 \pm 3.138 \\
\mu_3 - \mu_5: & (23.525 - 20.900) \pm 3.138 & = 2.625 \pm 3.138 \\
\mu_4 - \mu_5: & (23.700 - 20.900) \pm 3.138 & = 2.800 \pm 3.138
\end{array}$$

Note that only two confidence intervals are not include zero. We are 90% confident that in average, thread 2 is stronger than threads 1 and 5, but also are unable to differentiate between the mean strength of threads 1, 3, 4, and 5 or threads 2, 3 and 4.

12.37 There are four null hypotheses and we test each at the .05 level of significance.

To construct the analysis of variance table, we must first calculate the treatment, row, column, and replicate sums. $T_{...} = 7.8$, $n = 4$, $r = 2$ so $C = (7.8)^2/(2.4)^2 = 1.90125$. Since

$$T_{(1)} = 2.1, \quad T_{(2)} = 1.86, \quad T_{(3)} = 1.77, \quad T_{(4)} = 2.07$$

so that $\sum_{k=1}^{n} T_{(k)}^2 = 15.2874$. Therefore,

$$SS(Tr) = \frac{1}{8} \cdot 15.2874 - 1.90125 = .009675$$

Next,

$$T_{1..} = 1.91, \quad T_{2..} = 1.89, \quad T_{3..} = 2.05, \quad T_{4..} = 1.95$$

so that $\sum_{i=1}^{n} T_{i..}^2 = 15.2252$ and

$$SSR = \frac{1}{8} \cdot 15.2252 - 1.90125 = .0019$$

Since

$$T_{.1.} = 2.29, \quad T_{.2.} = 1.73, \quad T_{.3.} = 1.62, \quad T_{.4.} = 2.16$$

so that $\sum_{j=1}^{n} T_{.j.}^2 = 15.527$ and $SSC = .039625$. Also, $T_{..1} = 3.91$ and $T_{..2} = 3.89$ so that $\sum_{l=1}^{r} T_{..l}^2 = 30.4202$ and $SS(reps) = .0000125$. Finally,

$$\sum_{i=1}^{n} \sum_{j=1}^{n} \sum_{l=1}^{r} y_{ij(k)l}^2 = 1.9628$$

so that $SST = 1.9628 - 1.90125 = .06155$ and

$$SSE = .06155 - .009675 - .0019 - .039625 - .0000125 = .0103375$$

The analysis-of-variance table is

Source of variation	Degrees of freedom	Sum of squares	Mean square	F
Labs	3	0.009675	0.003225	6.551
Rolling dir.	3	0.0019	0.000633	1.286
Across dir.	3	0.039625	0.013208	26.829
Reps	1	0.0000125	0.0000125	0.025
Error	21	0.103375	0.0004923	
Total	31	0.06155		

At the .05 level of significance, we conclude that, we cannot reject the null hypothesis of no replication effect, or the null hypothesis that the mean tin-coating weight does not change along the rolling direction(rows). Since $F_{3,21} = 3.07$, we reject the null hypothesis that laboratories are consistent (Tr) and the null hypothesis that mean weight does not change across rolling direction(columns).

12.39 There are four null hypotheses and we test each at the .01 level of significance.

To construct the analysis of variance table, we evaluate the sums of squares. $C = T_{..}^2/n^2 = 7.78^2/5^2 = 2.421136$

$$SST = \sum_{i=1}^{n}\sum_{j=1}^{n} y_{ij}^2 - C = 2.7648 - 2.421136 = .343664$$

$$SSR = \frac{1}{n}\sum_{k=1}^{n} T_{i.}^2 - C = 2.4852 - 2.421136 = .064064$$

$$SSC = \frac{1}{n}\sum_{j=1}^{n} T_{.j}^2 - C = 2.68772 - 2.421136 = .266584$$

$$SS(Latin) = \frac{1}{n}\sum_{k=1}^{n} T_{(k)}^2 - C = 2.4216 - 2.421136 = .000464$$

$$SS(Greek) = \frac{1}{n}\sum_{l=1}^{n} T_{(l)}^2 - C = 2.4324 - 2.421136 = .011264$$

$$SSE = SST - SS(Latin) - SS(Greek) - SSR - SSC = .001288$$

The analysis-of-variance table is

Source of variation	Degrees of freedom	Sum of squares	Mean square	F
Restraints(Latin)	4	0.000464	0.000116	0.72
Size(Rows)	4	0.064064	0.016016	99.48
Speed(Cols)	4	0.266584	0.066646	413.95
Angle(Greek)	4	0.011264	0.002816	17.49
Error	8	0.001288	0.000161	
Total	24	0.343664		

Since $F_{4,8} = 7.01$ for $\alpha = .01$, we cannot reject the null hypothesis of no mean difference due to restraint system. However, auto size, impact speed and impact angle are all significant.

12.41 (a) A possible arrangement is

Blocks	Treatments
1	1 and 2
2	1 and 3
3	2 and 3

The minimum number of replicates(blocks) is 3 and the minimum number of times a treatment appears is 2.

(b) A possible arrangement is

Blocks	Treatments
1	1, 2 and 3
2	1, 2 and 4
3	1, 3 and 4
4	2, 3 and 4

The minimum number of replicates(blocks) is 4 and the minimum number of times a treatment appears is 3.

(c) A possible arrangement is

Blocks	Treatments	Blocks	Treatments
1	1, 2, 3, 4	9	1, 3, 5, 6
2	1, 2, 3, 5	10	1, 4, 5, 6
3	1, 2, 3, 6	11	2, 3, 4, 5
4	1, 2, 4, 5	12	2, 3, 4, 6
5	1, 2, 4, 6	13	2, 3, 5, 6
6	1, 2, 5, 6	14	2, 4, 5, 6
7	1, 3, 4, 5	15	3, 4, 5, 6
8	1, 3, 4, 6		

The minimum number of replicates(blocks) is 15 and the minimum number of times a treatment appears is 10. In general, the minimum number of replicates(blocks) is $n!/k!(n-k)!$ and the minimum number of times a treatment appears is $(n-1)!/(k-1)!(n-k)!$.

12.43 The null hypothesis is that track designs are equally resistant to breakage. The level is $\alpha = .01$. The analysis-of-covariance table is

Source of variation	Sum of squares for x	Sum of squares for y	Sum of products
Treatments	420.130	114.55	192.33
Error	556.052	122.40	237.74
Total	976.182	236.95	430.07

Source of variation	Sum of squares for y'	Degrees of freedom	Mean square
Treatments	26.723	3	8.908
Error	20.754	15	1.384
Total	47.477	18	

Thus, the F statistic is 6.44. Since the critical value at the 0.01 level for an F distribution with 3 and 15 degrees of freedom is 5.42, we reject the null hypothesis.

The effect of usage on breakage resistance, that is, the regression coefficient of the model, is given by $SPE/SSE_x = .428$.

12.45 The null hypothesis is that the weight losses for the three lubricants are the same. The alternative is that the weight losses are not the same. The analysis-of-variance table is

Source of variation	Degrees of freedom	Sum of squares	Mean square	F
Flows	3	0.528	0.1760	2.80
Error	16	1.004	0.0628	
Total	19	1.532		

Since the critical value at the 0.05 level for an F distribution with 3 and 16 degrees of freedom is 3.24, we cannot reject the null hypothesis that flow through the precipitator has no effect on the exit loading.

12.47 (a) We first find $a = 3$, $\bar{y}_{..} = 9$, $\bar{y}_{1.} = 7$, $\bar{y}_{2.} = 8$ and $\bar{y}_{3.} = 12$. And $b = 4$,

$\bar{y}_{.1} = 8$, $\bar{y}_{.2} = 13$, $\bar{y}_{.3} = 4$ and $\bar{y}_{.4} = 11$. Thus,

$$\begin{array}{ccc} \text{Obs.} & & \text{Grand mean} \\ y_{ij} & & \bar{y}_{..} \\ \begin{bmatrix} 9 & 10 & 2 & 7 \\ 6 & 3 & 1 & 12 \\ 9 & 16 & 9 & 14 \end{bmatrix} & = & \begin{bmatrix} 9 & 9 & 9 & 9 \\ 9 & 9 & 9 & 9 \\ 9 & 9 & 9 & 9 \end{bmatrix} \end{array}$$

$$\begin{array}{cccc} & \text{Tr. effect} & & \text{Bl. effect} \\ & \bar{y}_{i.} - \bar{y}_{..} & & \bar{y}_{.j} - \bar{y}_{..} \\ + & \begin{bmatrix} -2 & -2 & -2 & -2 \\ -1 & -1 & -1 & -1 \\ 3 & 3 & 3 & 3 \end{bmatrix} & + & \begin{bmatrix} -1 & 4 & -5 & 2 \\ -1 & 4 & -5 & 2 \\ -1 & 4 & -5 & 2 \end{bmatrix} \end{array}$$

$$\begin{array}{cc} & \text{Error} \\ & y_{ij} - \bar{y}_{i.} - \bar{y}_{.j} + \bar{y}_{..} \\ + & \begin{bmatrix} 3 & -1 & 0 & -2 \\ -1 & 1 & -2 & 2 \\ -2 & 0 & 2 & 0 \end{bmatrix} \end{array}$$

(b) The sums of squares and degrees of freedoms are

$$\begin{aligned} SS(Tr) &= 4(-2)^2 + 4(-1)^2 + 4(3)^2 = 56, \quad df = 2 \\ SS(Bl) &= 3(-1)^2 + 3(4)^2 + 3(-5)^2 + 3(2)^2 = 138, \quad df = 3 \\ SSE &= 3^2 + (-1)^2 + 0^2 + \cdots + 0^2 + 2^2 + 0^2 = 32, \\ & \qquad df = (3-1)(4-1) = 6 \\ SST &= 9^2 + 10^2 + 2^2 + \cdots + 16^2 + 9^2 + 14^2 - 12(9)^2 = 226, \\ & \qquad df = 11 \\ &= SS(Tr) + SS(Bl) + SSE = 226. \quad (check) \end{aligned}$$

(c) The null hypothesis is that the the four treatment population means are the same. The alternative is that they are not the same. A second null hypothesis is that there is no block effect. The analysis-of-variance table is

Source of variation	Degrees of freedom	Sum of squares	Mean square	F
Treatments	2	56	28.00	5.25
Blocks	3	138	46.00	8.63
Error	6	32	5.33	
Total	11	226		

Since the critical value at the 0.05 level for an F distribution with 2 and 6 degrees of freedom is 5.14, we can reject the null hypothesis. The treatment means are not the same. Since $F_{.05}$ with 3 and 6 degrees of freedom is 4.76, the block effect of the experiment is also apparent.

12.49 (a) The null hypotheses are (i) there is no fuel effect and (ii) there is no launcher effect. Each will be tested with $\alpha = .05$. The two-way analysis-of-variance table is

Source of variation	Degrees of freedom	Sum of squares	Mean square	F
Fuels	3	178.18	59.39	0.35
Launchers	1	0.45	0.45	0.0026
Error	3	513.53	171.18	
Total	7	692.16		

Since the critical value at the 0.05 level for an F distribution with 3 and 3 degrees of freedom is 9.28, we cannot reject the null hypothesis that there is no fuel effect.

(b) The null hypotheses is that there is no launcher effect. Since the critical value at the 0.05 level for an F distribution with 1 and 3 degrees of freedom is 10.10, we cannot reject the null hypothesis that there is no launcher effect.

12.51 The null hypotheses are (i) that there is no flow effect and (ii) that there is no precipitator effect. Each will be tested with $\alpha = .01$. Referring to Exercise 12.43, the two-way analysis-of-variance table is

Source of variation	Degrees of freedom	Sum of squares	Mean square	F
Flow	3	0.528	0.1760	7.75
Precipitators	4	0.732	0.1830	8.06
Error	12	0.272	0.0227	
Total	19	1.532		

Since the critical value at the 0.01 level for an F distribution with 3 and 12 degrees of freedom is 5.95, we reject the null hypothesis that there is no flow effect. Since the critical value at the 0.01 level for an F distribution with 4 and 12 degrees of freedom is 5.41, we also reject the null hypothesis that there is no precipitator effect.

The mean square error has dropped to 0.0227 from 0.0628 in the one-way analysis. This drop is to be expected since that variation due to precipitators is no longer pooled into the the mean square error.

12.53 The (sorted) sample means for the five sites are

E	B	C	A	D
3.70	6.77	14.20	21.30	26.73

and, from Exercise 12.48, the $MSE = 2.745$ with 8 degrees of freedom. From Table 12(a) with 8 degrees of freedom , after multiplying each r_p by $s_{\bar{y}} = \sqrt{MSE/n} = \sqrt{2.745/3} = 0.957$ to get R_p, we obtain

p	2	3	4	5
r_p	3.26	3.40	3.48	3.52
R_p	3.12	3.25	3.33	3.37

We conclude that, at the .05 level of significance, the means for all sites differ except site E and site B which have the lowest level of contamination. Site D has the highest level of contamination.

E	B	C	A	D
3.70	6.77	14.20	21.30	26.73

12.55 The analysis-of-variance table is

Source of variation	Degrees of freedom	Sum of squares	Mean square	F
Designs(Tr)	2	32853.41	16426.71	66.62
Pros(Rows)	2	12007.19	6003.60	24.35
Drivers(Cols)	2	1344.52	672.26	2.73
Fairways(Reps)	2	36140.52	18070.26	73.29
Error	18	4437.99	246.56	
Total	26	86783.63		

Since $F_{2,18} = 6.01$ for $\alpha = .01$, we reject the null hypothesis of no mean difference due to difference in design. We also reject the null hypothesis of no difference among pros and the null hypothesis of no difference among fairways. However, we cannot reject the null hypothesis of no difference among drivers.

We could use the Duncan multiple range test. The (sorted) average distances for the three golf ball designs are

Design C	Design B	Design A
210.56	253.33	296.00

From Table 12(a) with 18 degrees of freedom, after multiplying each r_p by $s_{\bar{y}} = \sqrt{MSE/n} = \sqrt{246.56/9} = 5.234$ to get R_p, we obtain

p	2	3
r_p	2.97	3.12
R_p	15.55	16.33

We conclude that, at the .05 level of significance, the average distance for design A is significantly higher than that of design C and B. designs. And the average distance for design C is significantly lower than design B and A.

Design C	Design B	Design A
210.56	253.33	296.00

12.57 To verify the analysis-of-variance table, we need to find the treatment sums of the sample.

$$T_{1.} = 7.25, \quad T_{2.} = 3.35, \quad T_{3.} = 7.10 \quad T_{4.} = 7.23.$$

Hence

$$T_{..} = 24.93 \quad \text{and} \quad C = \frac{24.93^2}{(4)(3)} = 51.7921.$$

Since $k = 4$, $n = 3$ and

$$\sum_{i=1}^{k} \sum_{j=1}^{n} y_{ij}^2 = 73.8769,$$

we have

$$SST = \sum_{i=1}^{k} \sum_{j=1}^{n} y_{ij}^2 - C^2 = 73.8769 - 51.7921 = 22.0848.$$

Now

$$\sum_{i=1}^{k} \frac{T_i^2}{n} = \frac{7.25^2}{3} + \frac{3.35^2}{3} + \frac{7.10^2}{3} + \frac{7.23^2}{3} = 55.4893.$$

Hence

$$SS(Tr) = \sum_{i=1}^{k} \frac{T_i^2}{n} - C = 55.4893 - 51.7921 = 3.6972$$

and

$$SSE = SST - SS(Tr) \;=\; 22.0848 - 3.6972 \;=\; 18.3876.$$

The F statistic is

$$F \;=\; \frac{MS(Tr)}{MSE} \;=\; \frac{3.6972/3}{18.3876/8} \;=\; .54.$$

Hence the analysis-of-variance table is

Source of variation	Degrees of freedom	Sum of squares	Mean square	F
Paper type	3	3.70	1.23	.54
Error	8	18.39	2.30	
Total	11	22.08		

which is the same as that shown in the text.

Chapter 13

FACTORIAL EXPERIMENTATION

13.1 1. We test each of the following hypotheses.

Null hypothesis (a): The replication effect is zero. $\rho_1 = \rho_2 = \rho_3 = 0$.

We reject the null hypothesis at .05 level if $F > F_{.05} = 4.84$ with 1 and 11 degrees of freedom.

Null hypothesis (b): The two-factor interaction is zero.

We reject the null hypothesis at .05 level if $F > F_{.05} = 2.93$ with 4 and 18 degrees of freedom.

Null hypothesis (c): The temperature(factor B) has no effect. $\beta_1 = \beta_2 = \beta_3 = \beta_4 = \beta_5 = 0$.

We reject the null hypothesis at .05 level if $F > F_{.05} = 2.93$ with 4 and 18 degrees of freedom.

Null hypothesis (d): The concentration(factor A) has no effect. $\alpha_1 = \alpha_2 = 0$.

We reject the null hypothesis at .05 level if $F > F_{.05} = 4.41$ with 1 and 18 degrees of freedom.

2. **Calculations:** The data are

Concentration grams/liter	Temperature (degrees F)	Reflectivity Rep 1	Rep 2	Rep 3	Sum
5	75	35	39	36	110
5	100	31	37	36	104
5	125	30	31	33	94
5	150	28	20	23	71
5	175	19	18	22	59
10	75	38	46	41	125
10	100	36	44	39	119
10	125	39	32	38	109
10	150	35	47	40	122
10	175	30	38	31	99
	Total	321	352	339	1,012

And the table of totals for the two factors are

	Factor B: Temperature(F^0) 75	100	125	150	175	
Factor A: Concentration 5	110	104	94	71	59	438
10	125	119	109	122	99	574
	235	223	203	193	158	1,012

Hence

$$C = \frac{1,012^2}{30} = 34,138.1333$$

And the sum of squares are

$$\begin{aligned}
SST &= \sum_i \sum_j \sum_k y_{ijk}^2 - C = 35,822 - C = 1,683.8667 \\
SS(Tr) &= \frac{1}{r} \sum_i \sum_j T_{ij.}^2 - C = \frac{1}{3}(110^2 + 104^2 + \cdots + 122^2 + 99^2) - C \\
&= 1,403.8667 \\
SSR &= \frac{1}{ab} \sum_k T_{..k}^2 - C = \frac{1}{(2)(5)}(312^2 + 352^2 + 339^2) - C = 48.4667 \\
SSE &= SST - SS(Tr) - SSR = 231.5333
\end{aligned}$$

$$\begin{aligned}
SSA &= \frac{1}{br}\sum_{i=1}^{a} T_{i..}^2 - C = \frac{1}{(5)(3)}(438^2 + 574^2) - C = 616.5334 \\
SSB &= \frac{1}{ar}\sum_{j=1}^{b} T_{.j.}^2 - C \\
&= \frac{1}{(2)(3)}(235^2 + 223^2 + 203^2 + 193^2 + 158^2) - C = 591.2000 \\
SS(AB) &= SS(Tr) - SSA - SSB = 196.1333
\end{aligned}$$

The analysis-of-variance table is

Source of variation	Degrees of freedom	Sums of squares	Mean square	F
Replication	2	48.4667	24.2334	1.88
A Concentration	1	616.5334	616.5334	47.93
B Temperature	4	591.2000	147.8000	11.49
AB interaction:	4	196.1333	49.0333	3.81
Error	18	231.5333	12.8630	
Total	29	1683.8667		

3. **Decision:** The replication effect is not significant at .05 level. The concentration effect and the temperature effect are both significant at .05 level. However, these cannot be interpreted individually because the interaction effect is significant at .05 level.

   The two-way table of fitted values, $\bar{y}_{ij.}$, provides a summary of the experiment.

		Factor B: Temperature(F^0)				
		75	100	125	150	175
Factor A:	5	36.67	34.67	31.33	23.67	19.67
Concentration	10	41.67	39.67	36.33	40.67	33.00

4. **Further Analysis:** The optimal reflectivity occurred at 75^0F with concentration 10 grams per liter. At optimal conditions, the reflectivity is a normal random variable with mean μ_{21} and variance σ^2. Since the mean square error(MSE) is an unbiased estimate for σ^2 and $t_{.025} = 2.101$ with 18 degrees of

freedom, the 95% confidence interval of reflectivity at optimal conditions is

$$\bar{y}_{21.} \pm t_{.025}\sqrt{\frac{MSE}{n}} = 41.67 \pm 2.101\sqrt{\frac{12.8630}{3}} = 41.67 \pm 4.35$$

or $37.32 < \mu_{21} < 46.02$.

13.3 The table of totals for the two factors are

		Engine 1	Engine 2	Engine 3	
Detergent	A	84	85	109	278
	B	91	92	100	283
	C	82	97	100	279
	D	89	82	106	277
		346	356	415	1117

We also have $a = 4$, $b = 3$, $r = 2$ and

$$T_{..1} = 565, \quad T_{..2} = 552, \quad \sum_{i=1}^{4}\sum_{j=1}^{3}\sum_{k=1}^{2} y_{ijk}^2 = 52,745$$

Hence $C = 1117^2/(4)(3)(2) = 51,987.04$.

$$\begin{aligned}
SST &= \sum_i\sum_j\sum_k y_{ijk}^2 - C = 52,745 - 51,987.04 = 757.96 \\
SS(Tr) &= \frac{1}{r}\sum_i\sum_j T_{ij.}^2 - C = \frac{1}{2}(84^2 + 85^2 + \cdots + 82^2 + 106^2) - 51,987.04 \\
&= 473.46 \\
SSR &= \frac{1}{ab}\sum_k T_{..k}^2 - C = \frac{1}{(4)(3)}(565^2 + 552^2) - 51,987.04 = 7.04 \\
SSE &= SST - SS(Tr) - SSR = 277.46 \\
SSA &= \frac{1}{br}\sum_{i=1}^{a} T_{i..}^2 - C = \frac{1}{(3)(2)}(278^2 + 283^2 + 279^2 + 277^2) - 51,987.04 \\
&= 3.46
\end{aligned}$$

$$\begin{aligned} SSB &= \frac{1}{ar}\sum_{j=1}^{b} T_{\cdot j \cdot}^2 - C = \frac{1}{(4)(2)}(346^2 + 356^2 + 415^2) - 51,987.04 \\ &= 347.59 \end{aligned}$$

$$SS(AB) = SS(Tr) - SSA - SSB = 122.41$$

The analysis-of-variance table is

Source of variation	Degrees of freedom	Sums of squares	Mean square	F
Replication	1	7.04	7.04	0.28
A (Detergent)	3	3.46	1.15	.05
B (Engine)	2	347.59	173.80	6.89
AB interaction:	6	122.41	20.40	0.81
Error	11	277.46	25.22	
Total	23	757.96		

Since $F_{.05} = 4.84$ with 1 and 11 degrees of freedom, the replication effect is not significant at the .05 level. Since $F_{.05} = 3.59$ with 3 and 11 degrees of freedom, the detergent effect is not significant at .05 level. Since $F_{.05} = 3.98$ with 2 and 11 degrees of freedom, the engine effect is significant at .05 level. Since $F_{.05} = 3.09$ with 6 and 11 degrees of freedom, the interaction effect is not significant at .05 level.

We conclude that the engine has an important effect. Machine 3 has a higher "cleanness" level than the other engines. All other effects may be unimportant.

13.5 The table of totals for the two factors are

		Filler 32	Filler 37	Filler 42	
Flow	3	4.09	4.64	5.17	13.90
Flow	12	4.63	4.61	5.13	14.37
Flow	30	4.65	5.11	5.20	14.96
		13.37	14.36	15.50	43.23

We also have $a = 3$, $b = 3$, $r = 2$ and

$$T_{..1} = 21.75, \quad T_{..2} = 21.48, \quad \sum_{i=1}^{3}\sum_{j=1}^{3}\sum_{k=1}^{2} y_{ijk}^2 = 104.4311$$

Hence

$$C = \frac{43.23^2}{(3)(3)(2)} = 103.82405.$$

$$\begin{aligned}
SST &= \sum_i\sum_j\sum_k y_{ijk}^2 - C = 104.43110 - 103.82405 = .60705 \\
SS(Tr) &= \frac{1}{r}\sum_i\sum_j T_{ij.}^2 - C \\
&= \frac{1}{2}(4.09^2 + 4.64^2 + \cdots + 5.11^2 + 5.20^2) - 103.82405 = .55950 \\
SSR &= \frac{1}{ab}\sum_k T_{..k}^2 - C \\
&= \frac{1}{(3)(3)}(21.75^2 + 21.48^2) - 103.82405 = .00405 \\
SSE &= SST - SS(Tr) - SSR = .04350 \\
SSA &= \frac{1}{br}\sum_{i=1}^{a} T_{i..}^2 - C = \frac{1}{(3)(2)}(13.90^2 + 14.37^2 + 14.96^2) - 103.82405 \\
&= .09403 \\
SSB &= \frac{1}{ar}\sum_{j=1}^{b} T_{.j.}^2 - C = \frac{1}{(3)(2)}(13.37^2 + 14.36^2 + 15.50^2) - 103.82405 \\
&= .37870 \\
SS(AB) &= SS(Tr) - SSA - SSB = .08677
\end{aligned}$$

The analysis-of-variance table is

Source of variation	Degrees of freedom	Sums of squares	Mean square	F
Replication	1	.00405	.00405	.74
A (Flow)	2	.09403	.04702	8.65
B (Filler)	2	.37870	.18935	34.82
AB interaction:	4	.08677	.02169	3.99
Error	8	.04350	.00544	
Total	17	.60705		

Since $F_{.05} = 5.32$ with 1 and 8 degrees of freedom, the replication effect is not significant at the .05 level. Since $F_{.01} = 8.65$ with 2 and 8 degrees of freedom, the Flow and Filler effects are both significant at .01 level. Since $F_{.05} = 3.84$ with 4 and 8 degrees of freedom, the Flow–Filter interaction effect is also significant at .05 level.

13.7 We have $a = 3$, $b = 2$, $c = 2$, $d = 3$, $r = 2$ and

$$T_{....1} = 1,059, \quad T_{....2} = 1,020, \quad T_{.....} = 2,079$$

$$\sum_{i=1}^{3}\sum_{j=1}^{2}\sum_{k=1}^{2}\sum_{m=1}^{3} T_{ijkm.}^2 = 127,787 \quad \text{and} \quad \sum_{i=1}^{3}\sum_{j=1}^{2}\sum_{k=1}^{2}\sum_{m=1}^{3}\sum_{l=1}^{2} y_{ijkml}^2 = 64,553$$

Hence $C = 2079^2/(3)(2)(2)(3)(2) = 60,031.125$.

$$\begin{aligned}
SST &= 64553.000 - 60031.125 = 4521.875 \\
SS(Tr) &= \frac{1}{2}(127787.000) - 60031.125 = 3862.375 \\
SSR &= \frac{1}{(3)(2)(2)(3)}\left(1059^2 + 1020^2\right) - 60031.125 = 21.125 \\
SSE &= SST - SS(Tr) - SSR = 4521.875 - 3862.375 - 21.125 \\
&= 638.375
\end{aligned}$$

Next we construct all possible two-way tables:

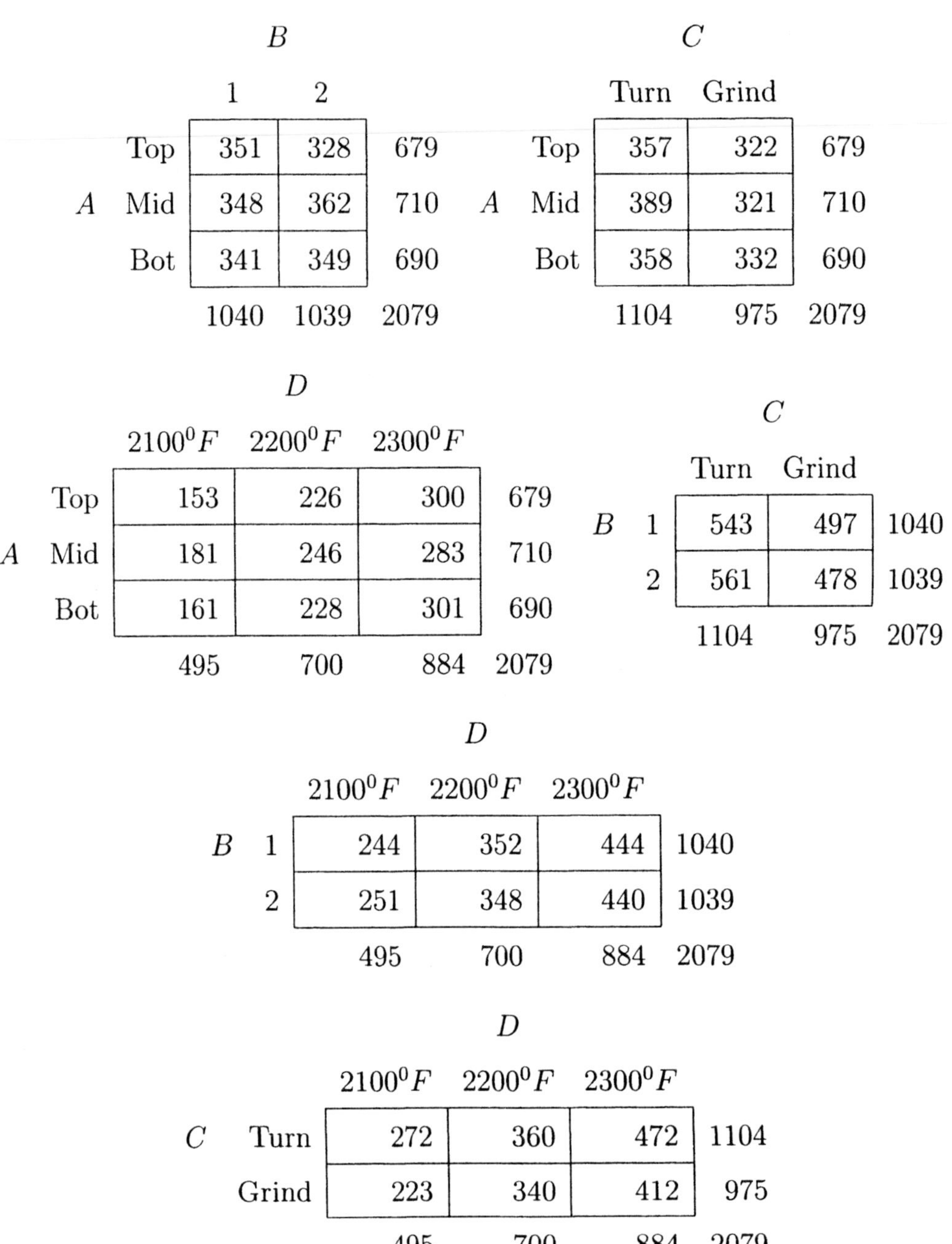

	B: 1	2	
A: Top	351	328	679
Mid	348	362	710
Bot	341	349	690
	1040	1039	2079

	C: Turn	Grind	
A: Top	357	322	679
Mid	389	321	710
Bot	358	332	690
	1104	975	2079

	D: 2100^0F	2200^0F	2300^0F	
A: Top	153	226	300	679
Mid	181	246	283	710
Bot	161	228	301	690
	495	700	884	2079

	C: Turn	Grind	
B: 1	543	497	1040
2	561	478	1039
	1104	975	2079

	D: 2100^0F	2200^0F	2300^0F	
B: 1	244	352	444	1040
2	251	348	440	1039
	495	700	884	2079

	D: 2100^0F	2200^0F	2300^0F	
C: Turn	272	360	472	1104
Grind	223	340	412	975
	495	700	884	2079

From the tables, we have

$$\begin{aligned} SSA &= \frac{1}{bcdr}\sum_{i=1}^{a} T_{i....}^2 - C = \frac{1,441,241}{(2)(2)(3)(2)} - 60,031.125 \\ &= 20.583 \\ SSB &= \frac{1}{acdr}\sum_{j=1}^{b} T_{.j...}^2 - C = \frac{2,161,121}{(3)(2)(3)(2)} - 60,031.125 \end{aligned}$$

$$
\begin{aligned}
&= .014 \\
SSC &= \frac{1}{abdr}\sum_{k=1}^{c} T^2_{..k..} - C = \frac{2,169,441}{(3)(2)(3)(2)} - 60,031.125 \\
&= 231.125 \\
SSD &= \frac{1}{abcr}\sum_{m=1}^{d} T^2_{...m.} - C = \frac{1,516,481}{(3)(2)(2)(2)} - 60,031.125 \\
&= 3,155.583
\end{aligned}
$$

Since

$$\frac{1}{cdr}\sum_{i=1}^{a}\sum_{j=1}^{b} T^2_{ij...} - C = \frac{721,015}{(2)(3)(2)} - 60,031.125 = 53.458$$

$$\frac{1}{bdr}\sum_{i=1}^{a}\sum_{k=1}^{c} T^2_{i.k..} - C = \frac{732,883}{(2)(3)(2)} - 60,031.125 = 292.458$$

$$\frac{1}{bcr}\sum_{i=1}^{a}\sum_{m=1}^{d} T^2_{i..m.} - C = \frac{506,357}{(2)(2)(2)} - 60,031.125 = 3,263.500$$

$$\frac{1}{adr}\sum_{j=1}^{b}\sum_{k=1}^{c} T^2_{.jk..} - C = \frac{1,085,063}{(3)(3)(2)} - 60,031.125 = 250.153$$

$$\frac{1}{acr}\sum_{j=1}^{b}\sum_{m=1}^{d} T^2_{.j.m.} - C = \frac{758,281}{(3)(2)(2)} - 60,031.125 = 3,158.960$$

$$\frac{1}{abr}\sum_{k=1}^{c}\sum_{m=1}^{d} T^2_{..km.} - C = \frac{761,441}{(3)(2)(2)} - 60,031.125 = 3,422.292$$

we have

$$
\begin{aligned}
SS(AB) &= 53.458 - 20.583 - .014 = 32.861 \\
SS(AC) &= 292.458 - 20.583 - 231.125 = 40.750 \\
SS(AD) &= 3,263.500 - 20.583 - 3,155.583 = 87.334 \\
SS(BC) &= 250.153 - .014 - 231.125 = 19.014 \\
SS(BD) &= 3,158.960 - .014 - 3,155.583 = 3.361 \\
SS(CD) &= 3,422.292 - 231.125 - 3,155.583 = 35.583
\end{aligned}
$$

Since

$$\frac{1}{dr}\sum_{i=1}^{a}\sum_{j=1}^{b}\sum_{k=1}^{c}T_{ijk..}^2 - C = \frac{362,313}{(3)(2)} - 60,031.125 = 354.375$$

$$\frac{1}{cr}\sum_{i=1}^{a}\sum_{j=1}^{b}\sum_{m=1}^{d}T_{ij.m.}^2 - C = \frac{253,477}{(2)(2)} - 60,031.125 = 3,338.125$$

$$\frac{1}{br}\sum_{i=1}^{a}\sum_{k=1}^{c}\sum_{k=1}^{d}T_{i.km.}^2 - C = \frac{254,921}{(2)(2)} - 60,031.125 = 3,699.125$$

$$\frac{1}{ar}\sum_{j=1}^{b}\sum_{k=1}^{c}\sum_{m=1}^{d}T_{.jkm.}^2 - C = \frac{380,913}{(3)(2)} - 60,031.125 = 3,454.375$$

we have

$$\begin{aligned}
SS(ABC) &= 354.375 - SSA - SSB - SSC - SS(AB) - SS(AC) \\
&\quad -SS(BC) = 10.028 \\
SS(ABD) &= 3,338.125 - SSA - SSB - SSD - SS(AB) - SS(AD) \\
&\quad -SS(BD) = 38.389 \\
SS(ACD) &= 3,699.125 - SSA - SSC - SSD - SS(AC) - SS(AD) \\
&\quad -SS(CD) = 128.167 \\
SS(BCD) &= 3,454.375 - SSB - SSC - SSD - SS(BC) - SS(BD) \\
&\quad -SS(CD) = 9.693 \\
SS(ABCD) &= SS(Tr) - SSA - SSB - SSC - SSD - SS(AB) \\
&\quad -SS(AC) - SS(AD) - SS(BC) - SS(BD) - SS(CD) \\
&\quad -SS(ABC) - SS(ABD) - SS(ACD) - SS(BCD) \\
&= 49.890
\end{aligned}$$

The analysis-of-variance table is

Source of variation	Degrees of freedom	Sums of squares	Mean square	F
Replication	1	21.125	21.125	1.158
Main effects:				
A	2	20.583	10.291	.564
B	1	.014	.014	.001
C	1	231.125	231.125	12.672
D	2	3,155.583	1,577.792	86.506
Two-factor interactions:				
AB	2	32.861	16.431	.901
AC	2	40.750	20.375	1.117
AD	4	87.334	21.834	1.197
BC	1	19.014	19.014	1.042
BD	2	3.361	1.805	.099
CD	2	35.583	17.792	.975
Three-factor interactions:				
ABC	2	10.028	5.014	.275
ABD	4	38.389	9.597	.526
ACD	4	128.167	32.042	1.757
BCD	2	9.693	4.847	.266
$ABCD$ interaction:	4	49.890	12.473	.684
Error	35	638.375	18.239	
Total	71	4,521.875		

At the .05 level, the critical values are

Degrees of freedom	F
1 and 35	4.12
2 and 35	3.27
4 and 35	2.64

Thus main effects C(specimen preparation) and D(twisting temperature) are significant at the .05 level. No other effects are significant at the .05 level.

From Figure 13.1, it is clear that the smallest number of turns to break the steel occurred at $2,100^0F$ for steel that had undergone grinding.

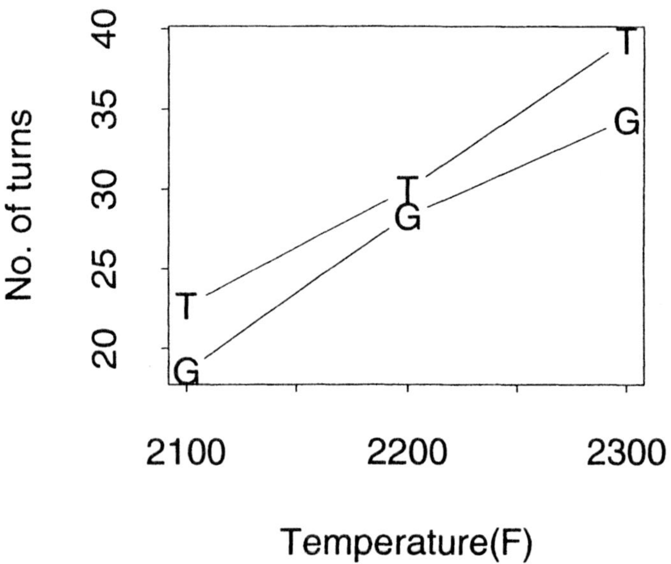

Figure 13.1: T and G stand for Turn and Grind, respectively. Exercise 13.7.

13.9 The four equations are:

$$\mu_{111} = \mu + \alpha_1 + \beta_1 + (\alpha\beta)_{11}$$
$$\mu_{121} = \mu + \alpha_1 - \beta_1 - (\alpha\beta)_{11}$$
$$\mu_{211} = \mu - \alpha_1 + \beta_1 - (\alpha\beta)_{11}$$
$$\mu_{221} = \mu - \alpha_1 - \beta_1 + (\alpha\beta)_{11}$$

Now solve the equations by combining the four population means on the left hand sides. We first sum the means.

$$\tfrac{1}{4}(\mu_{111} + \mu_{121} + \mu_{211} + \mu_{221}) = \mu$$
$$\tfrac{1}{4}(\mu_{111} + \mu_{121} - \mu_{211} - \mu_{221}) = \alpha_1$$
$$\tfrac{1}{4}(\mu_{111} - \mu_{121} + \mu_{211} - \mu_{221}) = \beta_1$$
$$\tfrac{1}{4}(\mu_{111} - \mu_{121} - \mu_{211} + \mu_{221}) = (\alpha\beta)_{11}$$

13.11 (a) The analysis-of-variance table for the two-way classification with 7 degrees of freedom for treatments and 1 degree of freedom for replicates is

Source of variation	Degrees of freedom	Sums of squares	Mean square	F
Replication	1	1.5625	1.5625	2.22
Treatments	7	103.9375	14.8482	21.05
Error	7	4.9375	.7054	
Total	15	110.4375		

Since $F_{.05} = 5.59$ with 1 and 7 degrees of freedom, the replication effect is not significant at the .05 level. Since $F_{.01} = 6.99$ with 7 and 7 degrees of freedom, the treatment effect is significant at the .01 level.

(b) We can use the table of signs on page 458. The data written in standard order are

	A Lubricant	B Heat	C Resin	Ratings Rep. 1	Rep. 2	Total
1	Fresh	Unheated	A	6	8	14
a	Aged	Unheated	A	6	7	13
b	Fresh	Heated	A	9	9	18
ab	Aged	Heated	A	9	8	17
c	Fresh	Unheated	B	8	7	15
ac	Aged	Unheated	B	6	8	14
bc	Fresh	Heated	B	1	2	3
abc	Aged	Heated	B	2	3	5

Thus, from the table of signs, the effect totals are

$$
\begin{aligned}
[I] &= 14+13+18+17+15+14+3+5 = 99 \\
[A] &= -14+13-18+17-15+14-3+5 = -1 \\
[B] &= -14-13+18+17-15-14+3+5 = -13 \\
[AB] &= 14-13-18+17+15-14-3+5 = 3 \\
[C] &= -14-13-18-17+15+14+3+5 = -25 \\
[AC] &= 14-13+18-17-15+14-3+5 = 3 \\
[BC] &= 14+13-18-17-15-14+3+5 = -29 \\
[ABC] &= -14+13+18-17+15-14-3+5 = 3
\end{aligned}
$$

(c) Since $n = 3$ and $r = 2$, we have

$$
\begin{aligned}
SSA &= (-1)^2/2(2^2) = .0625 & SSB &= (-13)^2/2(2^2) = 10.0625 \\
SSC &= (-25)^2/2(2^2) = 39.0625 & SSAB &= (3)^2/2(2^2) = .5625 \\
SSAC &= (3)^2/2(2^2) = .5625 & SSBC &= (-29)^2/2(2^2) = 52.5625 \\
SSABC &= (3)^2/2(2^2) = .5625
\end{aligned}
$$

To check that these sums of squares add up to $SS(Tr)$ from part (a), we have

$$.0625 + 10.5625 + 39.0625 + .5625 + .5625 + 52.5625 + .5625 = 103.9375$$

as required.

(d) The data were arranged in standard order in part (b). The table of calculations using the Yates method is

Exp. con.	Tr. total	1	2	3	Sum of squares
1	14	27	62	99	612.5625
a	13	35	37	−1	.0625
b	18	29	−2	−13	10.5625
ab	17	8	1	3	.5625
c	15	−1	8	25	39.0625
ac	14	−1	−21	3	.5625
bc	3	−1	0	−29	52.5625
abc	5	2	3	3	.5625

These numbers match those of parts (b) and (c).

The analysis-of-variance table is

Source of variation	Degrees of freedom	Sums of squares	Mean square	F
Replication	1	1.5625	1.5625	2.2151
Main effects:				
A	1	.0625	.0625	.0886
B	1	10.5625	10.5625	14.9738
C	1	39.0625	39.0625	55.3764
Two-factor interactions:				
AB	1	.5625	.5625	.7974
AC	1	.5625	.5625	.7974
BC	1	52.5625	52.5625	74.5145
ABC interaction:	1	.5625	.5625	.7974
Error	7	4.9375	.7054	
Total	15	110.4375		

The critical value is $F_{.05} = 5.59$ with 1 and 7 degrees of freedom. The effects for B, C and BC interaction are significant at the .05 level. We don't interpret main effects individually because of the apparent BC interaction.

13.13 The table of calculations using the Yates method is given in the first table. The table for analysis-of-variance follows.

Yates method for Exercise 13.13						
Exp. con.	Tr. total	1	2	3	4	Sum of squares
1	82.2	157.7	337.3	690.6	1376.6	59,219.6113
a	75.5	179.6	353.3	686.0	−44.2	61.0513
b	98.4	166.5	352.5	−5.2	92.6	267.9612
ab	81.2	186.8	333.5	−39.0	25.8	20.8013
c	84.3	164.5	−23.9	42.2	−3.0	.2813
ac	82.2	188.0	18.7	50.4	58.2	105.8513
bc	83.0	153.3	−27.3	12.4	1.8	.1013
abc	103.8	180.2	−11.7	13.4	76.6	183.3612
d	82.0	−6.7	21.9	16.0	−4.6	.6613
ad	82.5	−17.2	20.3	−19.0	−33.8	35.7013
bd	107.9	−2.1	23.5	42.6	8.2	2.1013
abd	80.1	20.8	26.9	15.6	1.0	.0313
cd	83.3	.5	−10.5	−1.6	−35.0	38.2813
acd	70.0	−27.8	22.9	3.4	−27.0	22.7813
bcd	89.3	−13.3	−28.3	33.4	5.0	.7813
$abcd$	90.9	1.6	14.9	43.2	9.8	3.0013

The analysis-of-variance table is

Source of variation	Degrees of freedom	Sums of squares	Mean square	F
Replication	1	300.1250	300.1250	.0046
Main effects:				
A	1	61.0513	61.0513	1.8468
B	1	267.9612	267.9612	8.1057
C	1	.2813	.2813	.0085
D	1	.6613	.6613	.0200
Two-factor interactions:				
AB	1	20.8013	20.8013	.6292
AC	1	105.8513	105.8513	3.2020
AD	1	35.7013	35.7013	1.0799
BC	1	.1013	.1013	.0030
BD	1	2.1013	2.1013	.0636
CD	1	38.2813	38.2813	1.1580

Three-factor interactions:				
ABC	1	183.3612	183.3612	5.5466
ABD	1	.0313	.0313	.0091
ACD	1	22.7813	22.7813	.6891
BCD	1	.7813	.7813	.0236
$ABCD$ interaction:	1	3.0013	3.0013	.0908
Error	15	495.8750	33.0583	
Total	31	1,538.7487		

The critical value is $F_{.05} = 4.54$ with 1 and 15 degrees of freedom. The replication effect is significant at the .05 level and so are the main effect of B and the ABC interaction. It seems that aging time has no effect on the gain of the semiconductor device.

13.15 From the model

$$y_{100l} = \mu + \alpha_1 + \beta_0 + \gamma_0 + (\alpha\beta)_{10} + (\alpha\gamma)_{10} + (\beta\gamma)_{00} + (\alpha\beta\gamma)_{100} + \rho_l + \epsilon_{100l}$$

Thus,

$$\sum_{l=1}^{r} y_{100l} = r\,(\,\mu + \alpha_1 + \beta_0 + \gamma_0 + (\alpha\beta)_{10} + (\alpha\gamma)_{10}$$

$$+(\beta\gamma)_{00} + (\alpha\beta\gamma)_{100}\,) + \sum_{l=1}^{r} \rho_l + \sum_{l=1}^{r} \epsilon_{100l}$$

Since $\sum_{l=1}^{r} \rho_l = 0$ and $(a) = \sum_{l=1}^{r} y_{100l}$, we have

$$(a) = r[\mu + \alpha_1 + \beta_0 + \gamma_0 + (\alpha\beta)_{10} + (\alpha\gamma)_{10}$$

$$+(\beta\gamma)_{00} + (\alpha\beta\gamma)_{100}] + \sum_{l=1}^{r} \epsilon_{100l}$$

But, the constraints are $\alpha_1 = -\alpha_0$,

$$(\alpha\beta)_{10} = -(\alpha\beta)_{00}\,, \quad (\alpha\gamma)_{10} = -(\alpha\gamma)_{00}\,, \quad (\alpha\beta\gamma)_{100} = -(\alpha\beta\gamma)_{000}$$

so that

$$(a) = r[\mu - \alpha_0 + \beta_0 + \gamma_0 - (\alpha\beta)_{00} - (\alpha\gamma)_{00} + (\beta\gamma)_{00} - (\alpha\beta\gamma)_{000}] + \sum_{l=1}^{r} \epsilon_{100l}$$

13.17 Multiplying each of the equations in the previous exercise by the appropriate sign and adding gives

$$-(1) + (a) - (b) + (ab) - (c) + (ac) - (bc) + (abc) = -8r\alpha_0 + \epsilon_A$$

where

$$\epsilon_A = \sum_{l=1}^{r} (-\epsilon_{000l} + \epsilon_{100l} - \epsilon_{010l} + \epsilon_{110l} - \epsilon_{001l} + \epsilon_{101l} - \epsilon_{101l} + \epsilon_{111l})$$

and

$$(1) + (a) - (b) - (ab) - (c) - (ac) + (bc) + (abc) = 8r(\beta\gamma)_{00} + \epsilon_{BC}$$

where

$$\epsilon_{BC} = \sum_{l=1}^{r} (\epsilon_{000l} + \epsilon_{100l} - \epsilon_{010l} - \epsilon_{110l} - \epsilon_{001l} - \epsilon_{101l} + \epsilon_{101l} + \epsilon_{111l})$$

13.19 Summing up the various sum of squares, we have

$$134,551 + 4,632 + 225,624 + 713 + 39,410 + 18,673 + 31,689 + 1,001$$

$$+24,698 + 81 + 39,130 + 30 + 12,601 + 14,070 + 385 = 547,228$$

This is the same as $SS(Tr)$ obtained in the calulations leading to the table.

13.21 (a) We verify the calculation of mean square using the formulas based on totals.

$$T_{1.} = 24, T_{2.} = 40, T_{3.} = 32, T_{4.} = 44, \text{ and } T_{..} = 40$$

so $C = 140^2/8 = 2,450$. Further

$$T_{.1} = 68, T_{.2} = 72, \text{ and } \sum\sum y_{ij}^2 = 2,588$$

so

$$\begin{aligned} SST &= 2,588 - 2,450 = 138 \\ SS(Tr) &= \frac{1}{2}(5,136) - 2,450 = 118 \end{aligned}$$

Therefore $SSE = 138 - 118 = 20$ with 4 degrees of freedom and $MSE = 20/4 = 5$ which verifies the result in the example.

(b)

$$\frac{[B]}{2r} = \frac{1}{2r}[-(1) - (a) + (b) + (ab)]$$

$$= \frac{1}{2r}[-\sum_{k=1}^{r} y_{00k} - \sum_{k=1}^{r} y_{10k} + \sum_{k=1}^{r} y_{01k} + \sum_{k=1}^{r} y_{11k}]$$

$$= \frac{1}{2r}[-\sum_{i=0}^{1}\sum_{k=1}^{r} y_{i0k} + \sum_{i=0}^{1}\sum_{k=1}^{r} y_{i1k}] = \bar{y}_{.1.} - \bar{y}_{.0.}$$

(c)

$$\frac{[AB]}{2r} = \frac{1}{2r}[(1) - (a) - (b) + (ab)]$$

$$= \frac{1}{2r}[\sum_{k=1}^{r} y_{00k} - \sum_{k=1}^{r} y_{10k} - \sum_{k=1}^{r} y_{01k} + \sum_{k=1}^{r} y_{11k}]$$

$$= \frac{1}{2}[\frac{1}{r}\sum_{k=1}^{r} y_{00k} - \frac{1}{r}\sum_{k=1}^{r} y_{10k} - \frac{1}{r}\sum_{k=1}^{r} y_{01k} + \frac{1}{r}\sum_{k=1}^{r} y_{11k}]$$

$$= \frac{1}{2}(\bar{y}_{11.} + \bar{y}_{00.}) - \frac{1}{2}(\bar{y}_{10.} + \bar{y}_{01.})$$

13.23 The visual summary of the four treatment means is given in Figure 13.2. To obtain

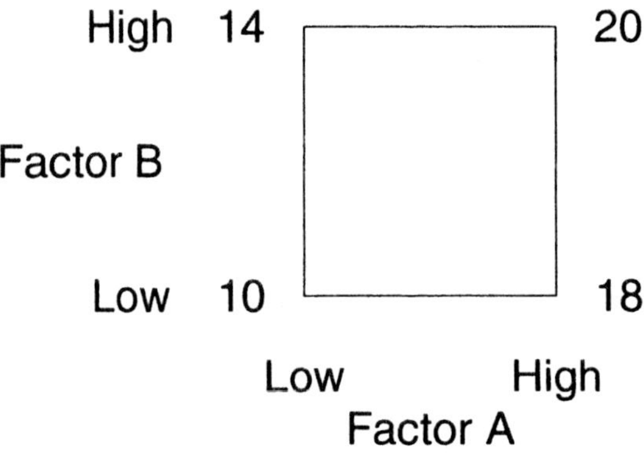

Figure 13.2: Visual summary of the experiment. Exercise 13.23

confidence intervals for the mean effects, we calculate

$$\widehat{\alpha_1 - \alpha_0} = \bar{y}_{1..} - \bar{y}_{0..} = 19 - 12 = 7$$

$$\widehat{\beta_1 - \beta_0} = \bar{y}_{.1.} - \bar{y}_{.0.} = 17 - 14 = 3$$

$$\widehat{\text{interaction}} = \frac{1}{2}(\bar{y}_{11.} - \bar{y}_{10.} - \bar{y}_{01.} + \bar{y}_{00.})$$

$$= \frac{1}{2}(20 - 18 - 14 + 10) = -1$$

Next, we find the mean square error.

$$\begin{aligned} C &= 124^2/8 = 1,922 \\ SST &= 2,140 - 1,922 = 218 \\ SS(Tr) &= \tfrac{1}{2}(4,080) - 1,922 = 118 \end{aligned}$$

so $SSE = 100$ with 4 degrees of freedom and $s^2 = MSE = 100/4 = 25$. Because $t_{.025} = 2.776$ for 4 degrees of freedom, the confidence intervals are

Factor A :

$$\bar{y}_{1..} - \bar{y}_{0..} \pm t_{.025}\sqrt{\frac{s^2}{r}} = 7 \pm (2.776)\sqrt{\frac{25}{2}} = 7 \pm 9.8 \text{ or } -2.8 \text{ to } 16.8$$

Factor B :

$$\bar{y}_{.1.} - \bar{y}_{.0.} \pm t_{.025}\sqrt{\frac{s^2}{r}} = 7 \pm (2.776)\sqrt{\frac{25}{2}} = 3 \pm 9.8 \text{ or } -6.8 \text{ to } 12.8$$

Interaction AB :

$$-1 \pm 9.8 \text{ or } -10.8 \text{ to } 8.8$$

All of the intervals cover 0 so none of the effects are significant. The experiment can, ultimately, be summarized by the common mean $\bar{y}_{...} = 124/8 = 15.5$ and $s = 5$.

13.25 (a) The data and replicate means are

	Rep. 1	Rep. 2	Mean	$y_{ijk1} - \bar{y}_{ijk.}$	$y_{ijk2} - \bar{y}_{ijk.}$
1	4.5	4.1	4.3	.2	−.2
a	3.8	3.4	3.6	.2	−.2
b	3.1	4.3	3.7	−.6	.6
ab	7.2	6.8	7.0	.2	−.2
c	5.4	5.0	5.2	.2	−.2
ac	4.5	4.9	4.7	−.2	.2
bc	4.2	5.4	4.8	−.6	.6
abc	7.3	6.9	7.1	.2	−.2

Note that $y_{ijk1} - \bar{y}_{ijk.} = -(y_{ijk2} - \bar{y}_{ijk.})$. Squaring the entries in the last two columns and summing, or taking twice the sum of squares for one column, we obtain

$$SSE = 2(.2^2 + .2^2 + \cdots + .2^2) - 2^3[(-.05)^2 + (.05)^2] = 1.92 - .04 = 1.88$$

Therefore, $s^2 = 1.88/7 = .2686$

(b) Since $t_{.025} = 2.365$ for 7 degrees of freedom, the half length of the confidence intervals is

$$t_{.025}\sqrt{\frac{s^2}{2r}} = 2.365\sqrt{\frac{.2686}{4}} = 0.61.$$

Replacing the half length of the confidence intervals shown in the example by .61, the resulting individual 95% confidence intervals are:

rate effect: $1.1 \pm .61$ or $.49$ to 1.71,

additive effect: $1.2 \pm .61$ or $.59$ to 1.81,

nozzle effect: $.8 \pm .61$ or $.19$ to 1.41,

rate $\times$ additive interaction: $1.7 \pm .61$ or 1.09 to 2.31,

rate $\times$ nozzle interaction: $-.2 \pm .61$ or $-.81$ to $.41$,

additive $\times$ nozzle interaction: $-.2 \pm .61$ or $-.81$ to $.41$,

rate $\times$ additive $\times$ nozzle interaction: $-.3 \pm .61$ or $-.91$ to $.31$.

13.27 The visual summary of the eight treatment means is given in Figure 13.3. According to visual procedure, we assume that there is no replication effects in the model. The error sum of squares is

$$SSE = \sum_{i=0}^{1}\sum_{j=0}^{1}\sum_{k=0}^{1}\sum_{l=1}^{r}(y_{ijkl} - \bar{y}_{ijk.})^2$$

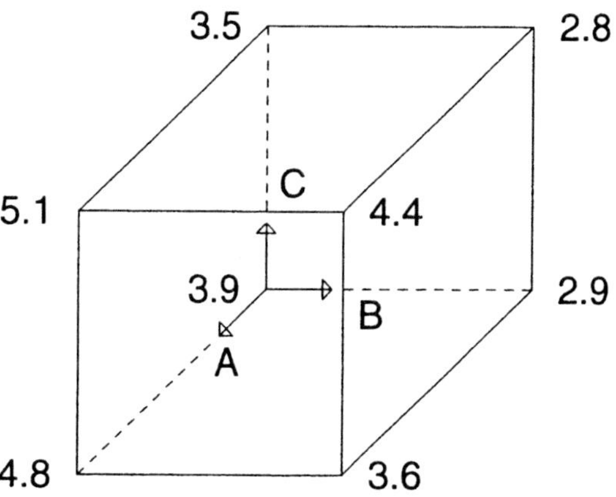

Figure 13.3: Visual summary of the experiment. Exercise 13.27

with $2^3(r-1)$ degrees of freedom. Thus we calculate

	Rep. 1	Rep. 2	Mean	$y_{ijk1} - \bar{y}_{ijk.}$	$y_{ijk2} - \bar{y}_{ijk.}$
1	3.7	4.1	3.9	$-.2$	$.2$
a	4.6	5.0	4.8	$-.2$	$.2$
b	3.1	2.7	2.9	$.2$	$-.2$
ab	3.4	3.8	3.6	$-.2$	$.2$
c	3.4	3.6	3.5	$-.1$	$.1$
ac	5.3	4.9	5.1	$.2$	$-.2$
bc	2.4	3.2	2.8	$-.4$	$-.4$
abc	4.7	4.1	4.4	$.3$	$-.3$
Mean	3.825	3.925	3.875		

Note that $y_{ijk1}-\bar{y}_{ijk.} = -(y_{ijk2}-\bar{y}_{ijk.})$. Squaring the entries in the last two columns and summing, or taking twice the sum of squares for one column, we obtain

$$SSE = 2((-.2)^2 + (-.2)^2 + \cdots + .3^2) = .92$$

with $2^3(2-1) = 8$ degrees of freedom. Therefore, $s^2 = MSE = 92/8 = .115$. Since $t_{.025} = 2.306$ for 8 degrees of freedom, the half length of the confidence intervals is

$$t_{.025}\sqrt{\frac{s^2}{2r}} = 2.306\sqrt{\frac{.115}{4}} = .39$$

The estimated effects are:

Factor A :

$$\bar{y}_{1...} - \bar{y}_{0...} = 4.475 - 3.275 = 1.20$$

Factor B :

$$\bar{y}_{.1..} - \bar{y}_{.0..} = 3.425 - 4.325 = -.90$$

Factor C :

$$\bar{y}_{..1.} - \bar{y}_{..0.} = 3.950 - 3.800 = .150$$

AB :

$$\frac{1}{2}(\bar{y}_{11..} - \bar{y}_{10..} - \bar{y}_{01..} + \bar{y}_{00..}) = \frac{1}{2}(4.0 - 4.95 - 2.85 + 3.7) = -.05$$

AC :

$$\frac{1}{2}(\bar{y}_{1.1.} - \bar{y}_{1.0.} - \bar{y}_{0.1.} + \bar{y}_{0.0.}) = \frac{1}{2}(4.75 - 4.2 - 3.15 + 3.4) = .4$$

BC :

$$\frac{1}{2}(\bar{y}_{.11.} - \bar{y}_{.10.} - \bar{y}_{.01.} + \bar{y}_{.00.}) = \frac{1}{2}(3.6 - 3.25 - 4.3 + 4.35) = .2$$

ABC :

$$\frac{1}{4}(\bar{y}_{111.} - \bar{y}_{101.} - \bar{y}_{011.} + \bar{y}_{001.}) - \frac{1}{4}(\bar{y}_{110.} - \bar{y}_{100.} - \bar{y}_{010.} + \bar{y}_{000.})$$

$$= \frac{1}{4}(4.4 - 5.1 - 2.8 + 3.5) - \frac{1}{4}(3.6 - 4.8 - 2.9 + 3.9) = .05$$

so the confidence intervals are:

(Viscosity) A :

$$1.2 \pm .39 \quad \text{or} \quad .81 \text{ to } 1.59$$

(Temperature) B :

$$-.9 \pm .39 \quad \text{or} \quad -1.29 \text{ to } -.51$$

(Additive) C :

$$.15 \pm .39 \quad \text{or} \quad -.24 \text{ to } .54$$

AB :

$$-.05 \pm .39 \quad \text{or} \quad -.44 \text{ to } .34$$

AC :

$$.4 \pm .39 \quad \text{or} \quad .01 \text{ to } .79$$

BC :

$$.2 \pm .39 \quad \text{or} \quad -.19 \text{ to } .59$$

ABC :

$$.05 \pm .39 \quad \text{or} \quad -.34 \text{ to } .44$$

Only the confidence intervals for the main effects of oil viscosity and temperature and the viscosity $\times$ additive interaction fail to cover 0. Over the conditions of this experiment, low temperature produces a higher mean for the response.

13.29 Similar to Exercise 13.21, it can be shown that in a 2^n design, for each treatment,

$$\text{treatment total} = (\text{estimated effect}) \times r2^{n-1}$$

Thus the sum of squares of a treatment is

$$\begin{aligned} \text{sum of squares} &= \frac{(\text{treatment total})^2}{r2^n} \\ &= \frac{[(\text{estimated effect}) \times r2^{n-1}]^2}{r2^n} \end{aligned}$$

$$= (\text{estimated effect})^2 \times r2^{n-2}$$

Since $n = 3$ and $r = 2$, using the estimated effects given in the example, we calculate

Treatment	Estimated effect	Sum of squares
A	1.1	$1.1^2(2)(2) = 4.84$
B	1.2	$1.2^2(2)(2) = 5.76$
C	.8	$.8^2(2)(2) = 2.56$
AB	1.7	$1.7^2(2)(2) = 11.56$
AC	$-.2$	$(-.2)^2(2)(2) = .16$
BC	$-.2$	$(-.2)^2(2)(2) = .16$
ABC	$-.3$	$(-.3)^2(2)(2) = .36$

From the example, we also know that $MSE = s^2 = .24$ with $2^3(2-1) = 8$ degrees of freedom. Hence the analysis-of-variance table is

Source of variation	Degrees of freedom	Sums of squares	Mean square	F
Main effects:				
A Rate	1	4.84	4.84	20.17
B Additive	1	5.76	5.76	24.00
C Nozzle	1	2.56	2.56	10.67
Two-factor interactions:				
AB	1	11.56	11.56	48.17
AC	1	.16	.16	.67
BC	1	.16	.16	.67
ABC interaction:	1	.36	.36	1.50
Error	8	1.92	.24	
Total	15	27.32		

Since $F_{.01} = 11.26$ with 1 and 8 degrees of freedom, the rate and additive effects and their interaction are significant at the .01 level. Since $F_{.05} = 5.32$ with 1 and 8 degrees of freedom, the nozzle effect is significant at the .05 level. No other effects are significant at the .05 level.

13.31 (a) Since,

$$(a-1)(b-1)(c-1)(d-1)$$

$$= abcd - acd - bcd + cd - abd + ad + bd - d$$

$$-abc + ac + bc - c + ab - a - b + 1$$

we can use the blocks

$$Block\ 1 : 1, ab, ac, bc, bd, ad, cd, abcd$$

$$Block\ 2 : a, b, c, abc, d, abd, bcd, acd$$

(b) Since

$$(a-1)(b-1)(c-1)(d+1)$$

$$= abcd - acd - bcd + cd - abd + ad + bd - d$$

$$+abc - ac - bc + c - ab + a + b - 1$$

we first divide into two blocks

$$Block\ 1 : 1, ab, ac, bc, d, abd, bcd, acd$$

$$Block\ 2 : a, b, c, abc, bd, ad, cd, abcd$$

Since

$$(a+1)(b-1)(c-1)(d-1)$$

$$= abcd - acd + bcd - cd - abd + ad - bd + d$$

$$-abc + ac - bc + c + ab - a + b - 1$$

we divide the two blocks above to get the required 4 blocks.

$$Block\ 1 : 1, bc, abd, acd$$

$$Block\ 2 : ab, ac, d, bcd$$

$$Block\ 3 : a, bd, cd, abc$$

$$Block\ 4 : b, ad, c, abcd$$

13.33 We first combine the data as if there are no confounding effects presented with the two replicates and calculate the treatment sum of squares using the Yates method.

Yates method for Exercise 13.33						
Exp. con.	Tr. total	1	2	3	4	Sum of squares
1	4.6	10.0	22.8	52.6	101.0	312.7813
a	5.4	12.8	29.8	48.4	3.0	.2813
b	5.6	13.2	19.0	6.2	11.4	4.0613
ab	7.2	16.6	29.4	−3.2	−5.8	1.0513
c	4.2	8.2	2.4	6.2	17.4	9.4613
ac	9.0	10.8	3.8	5.2	5.0	.7813
bc	8.8	13.4	−3.4	−5.0	.6	.0113
abc	7.8	16.0	.2	−.8	.2	.0013
d	4.0	.8	2.8	7.0	−4.2	.5513
ad	4.2	1.6	3.4	10.4	−9.4	2.7613
bd	7.2	4.8	2.6	1.4	−1.0	.0313
abd	3.6	−1.0	2.6	3.6	4.2	.5513
cd	7.4	.2	.8	.6	3.4	.3613
acd	6.0	−3.6	−5.8	0.0	2.2	.1513
bcd	7.2	−1.4	−3.8	−6.6	−.6	.0113
$abcd$	8.8	1.6	3.0	6.8	13.4	5.6113

The table of block sums is

	Block 1	Block 2
Trial 1	28.8	28.4
Trial 2	23.2	20.6

Since $C = 318.7813$, we have

$$SST = 349.6400 - 318.7813 = 30.8588$$

$$SS(Tr) = 688.9200/2 - 318.7813 = 25.6788$$

$$SS(Bl) = 2598.6000/8 - 318.7813 = 6.0438$$

The analysis-of-variance table is

Source of variation	Degrees of freedom	Sums of squares	Mean square	F
Blocks	3	6.04375	2.0146	5.94
Main effects:				
A	1	.2813	.2813	.83
B	1	4.0613	4.0613	11.98
C	1	9.4613	9.4613	27.90
D	1	.5513	.5513	1.63
Two-factor interactions:				
AB	1	1.0513	1.0513	3.10
AC	1	.7813	.7813	2.30
AD	1	2.7613	2.7613	8.14
BC	1	.0113	.0113	.03
BD	1	.0313	.0313	.09
CD	1	.3613	.3613	1.07
Three-factor interactions:				
ABC	1	.0013	.0013	.00
ABD	1	.5513	.5513	1.63
ACD	1	.1513	.1513	.45
BCD	1	.0113	.0113	.03
Intrablock error	14	4.7475	.3391	
Total	31	30.8588		

The critical value is $F_{.05} = 4.60$ with 1 and 14 degrees of freedom. The effects for blocks, factors B and C and AD interaction are significant at the .05 level.

We summarize the findings in the following:

Tranq. B		Tranq. C		Tranq.
High	Low	High	Low	B High and C High
3.5	2.8	3.7	2.6	4.08

		Tranq. D	
		High	Low
Tranq. A	High	2.8	3.7
	Low	3.2	2.9

Running tranquilizers B and C at high doses level gives the best result since the effects of B and C are additive. It may also be advantagous to run A high simultaneously with D low.

13.35 The method of Exercise 13.20 allows the sign of any treatment total to be determined in the sum for any effect total. The odd-even method allows separation of the treatment totals according to sign. Thus, we most show that all "evens" have the same sign and all "odds" have the opposite sign. But the sign for a treatment total in the expansion of the appropriate product is -1 raised to the number of letters in the effect that are not in the treatment total. For example, for effect AB , the sign of (acd) is -1 since there is one letter (B) in AB that is not in (acd). Thus, "evens" must all have the same signs, and "odds" must have the opposite sign.

13.37 The experimental conditions are

$$1, ab, ac, ad, ae, af, bc, bd, be, bf, cd, ce, cf, de, df, ef,$$

$$abcd, abce, abcf, abde, abdf, abef, abde, acdf, acef,$$

$$adef, bcde, bcdf, bcef, bdef, cdef, abcdef,$$

The alias pairs are

		D and ABCEF	ABCE and DF
A and BCDEF	E and ABCDF	AD and BCEF	DE and ABCF
B and ACDEF	AE and BCDF	BD and ACEF	ADE and BCF
AB and CDEF	BE and ACDF	ABD and CEF	BDE and ACF
C and ABDEF	ABE and CDF	CD and ABEF	ABDE and CF
AC and BDEF	CE and ABDF	ACD and BEF	CDE and ABF
BC and ADEF	ACE and BDF	BCD and AEF	ACDE and BF
ABC and DEF	BCE and ADF	ABCD and EF	BCEE and AF
			ABCDE and F

13.39 (a) The modified standard order is:

$$1, af, bf, ab, cf, ac, bc, abcf, df, ad, bd, abdf, cd, acdf,$$

$$bcdf, aabcd, ef, ae, be, aabef, ce, acef, bcef, abce, de,$$

$$adef, bdef, abde, cdef, acde, bcde, abcdef.$$

(b) The procedure for a 2^n factorial experiment is the same as for the 2^6 case. First write out the 2^{n-1} treatment conditions for the first $n-1$ letters. Then, append the n-th letter to these 2^{n-1} treatment combinations as required to obtain the same treatments as in the half replicate block being used.

13.41 We are given that

$$[A] = (a) - (b) - (c) + (abc).$$

and, by Exercises 13.16 and 13.17 ,

$$(a) = r[\mu - \alpha_0 + \beta_0 + \gamma_0 - (\alpha\beta)_{00} - (\alpha\gamma)_{00}$$

$$+(\beta\gamma)_{00} - (\alpha\beta\gamma)_{000}] + \sum_{l=1}^{r} \epsilon_{100l}$$

$$(b) = r[\mu + \alpha_0 - \beta_0 + \gamma_0 - (\alpha\beta)_{00} + (\alpha\gamma)_{00}$$

$$-(\beta\gamma)_{00} - (\alpha\beta\gamma)_{000}] + \sum_{l=1}^{r} \epsilon_{010l}$$

$$(c) = r[\mu + \alpha_0 + \beta_0 - \gamma_0 + (\alpha\beta)_{00} - (\alpha\gamma)_{00}$$

$$-(\beta\gamma)_{00} - (\alpha\beta\gamma)_{000}] + \sum_{l=1}^{r} \epsilon_{001l}$$

$$(abc) = r[\mu - \alpha_0 - \beta_0 - \gamma_0 + (\alpha\beta)_{00} + (\alpha\gamma)_{00}$$

$$+(\beta\gamma)_{00} - (\alpha\beta\gamma)_{000}) + \sum_{l=1}^{r} \epsilon_{111l}$$

Thus,

$$[A] = r[-4\alpha_0 + 4(\beta\gamma)_{00}] + \sum_{l=1}^{r} (\epsilon_{100l} - \epsilon_{010l} - \epsilon_{001l} + \epsilon_{111l})$$

$$= -4r[\alpha_0 - (\beta\gamma)_{00}] + \epsilon_A$$

where

$$\epsilon_A = \sum_{l=1}^{r} (\epsilon_{100l} - \epsilon_{010l} - \epsilon_{001l} + \epsilon_{111l})$$

Since the expected values of $\epsilon_{100l}, \epsilon_{010l}$, ϵ_{001l} and ϵ_{111l} are zero, the expected value of ϵ_A is zero.

13.43 First, we arrange the data in modified standard order and calculate the following table:

Experimental conditions	Rep. 1	Rep 2.	Difference	$(y_1 - y_2)^2/2$
1	39.0	43.2	−4.2	8.82
ad	42.0			
bd	54.9			
ab	40.9	40.3	0.6	0.18
cd	43.1			
ac	29.3			
bc	34.8	48.2	−13.4	89.78
abcd	41.4	49.5	−8.1	32.805

The error sum of squares, obtained by summing the last column of the table, is

$$8.82 + 0.18 + 89.78 + 32.805 = 131.585$$

There are 4 degrees of freedom so the mean squared error is

$$\frac{131.585}{4} = 32.896$$

We use Yates methods to calculate effect totals. Similar to Exercise 13.38, we put the appended letter in parentheses to keep in mind that it is nothing to do with the effect totals calculated.

Yates method for Exercise 13.43					
Exp. con.	Tr. total	1	2	3	Sum of squares
1	39.0	81.0	176.8	325.4	13235.645
$a(d)$	42.0	95.8	148.6	−18.2	41.405
$b(d)$	54.9	72.4	−11.0	18.6	43.245
ab	40.9	76.2	−7.2	3.4	1.445
$c(d)$	43.1	3.0	14.8	−28.2	99.405
ac	29.3	−14.0	3.8	3.8	1.805
bc	34.8	−13.8	−17.0	−11.0	15.125
$abc(d)$	41.4	6.6	20.4	37.4	174.845

The analysis-of-variance table is

Source of variation	Degrees of freedom	Sums of squares	Mean square	F
Main effects:				
$A = BCD$	1	41.405	41.405	1.259
$B = ACD$	1	43.245	43.245	1.315
$C = ABD$	1	99.405	99.405	3.022
$D = ABC$	1	174.845	174.845	5.315
$AB = CD$	1	1.445	1.445	0.044
$AC = BD$	1	1.805	1.805	0.055
$BC = AD$	1	15.125	15.125	0.460
Error	4	131.585	32.896	
Total	11	508.860		

The critical value is $F_{.05} = 7.71$ with 1 and 4 degrees of freedom. None of the effects or interactions is significant at the 5 percent level .

13.45 (a) As a two-way classification with 12 treatments and three replicates, the analysis-of-variance table is

Source of variation	Degrees of freedom	Sums of squares	Mean square	F
Replication	2	.0039	.00195	.16
Treatments	11	12.7830	1.16209	97.25
Error	22	.2628	.01195	
Total	35	13.0497		

Since $F_{.05} = 3.44$ with 2 and 22 degrees of freedom, the replication effect is not significant at the .05 level. Since $F_{.05} = 2.26$ with 11 and 22 degrees of freedom, the treatment effect is significant at the .05 level.

(b) The three two-way tables:

		(*B*) Time(min.)		
		20	30	Totals
(*A*) Temp.	350^0F	12.7	15.2	27.9
	400^0F	14.6	16.8	31.4
	Totals	27.3	32.0	59.3

		(C) Tenderizer			
		A	B	C	Totals
(A) Temp.	350^0F	9.4	13.3	5.2	27.9
	400^0F	10.2	14.9	6.3	31.4
	Totals	19.6	28.2	11.5	59.3

		(C) Tenderizer			
		A	B	C	Totals
(B) Time	20	8.7	13.1	5.5	27.3
	30	10.9	15.1	6.0	32.0
	Totals	19.6	28.2	11.5	59.3

From the first table, we have

$$\begin{aligned}
SSA &= 1,764.37/(2\cdot 3\cdot 3) - 97.6803 = .3403 \\
SSB &= 1,769.29/(2\cdot 3\cdot 3) - 97.6803 = .6136 \\
SSC &= 1,311.65/(2\cdot 2\cdot 3) - 97.6803 = 11.6239 \\
SS(AB) &= 887.73/(3\cdot 3) - 97.6803 - SSA - SSB = .0025 \\
SS(AC) &= 658.03/(2\cdot 3) - 97.6803 - SSA - SSC = .0272 \\
SS(BC) &= 660.37/(2\cdot 3) - 97.6803 - SSB - SSC = .1439 \\
SS(ABC) &= SS(Tr) - SSA - SSB - SSC - SS(AB) - SS(AC) \\
&\quad -SS(BC) = .0316
\end{aligned}$$

(c) The analysis-of-variance table is

Source of variation	Degrees of freedom	Sums of squares	Mean square	F
Replication	2	.0039	.00195	.16
Main effects:				
A Temperature	1	.3403	.34030	28.48
B Time	1	.6136	.61360	51.35
C Tenderizer	2	11.6239	5.81195	486.36
Two-factor interactions:				
AB	1	.0025	.00250	.21
AC	2	.0272	.01360	1.14
BC	2	.1439	.07193	6.02
ABC interaction:	2	.0316	.01580	1.32
Error	22	.2628	.01195	
Total	35	13.0497		

Since $F_{.05} = 3.44$ with 2 and 22 degrees of freedom, the tenderizer effect and time-tenderizer interaction are significant at the .05 level. Since $F_{.05} = 4.30$ with 1 and 22 degrees of freedom, the main effects of temperature and of time are significant at the .05 level. No other effects are significant at the .05 level.

13.47 The table of $\bar{y}_{ij.}$ for the two factors are

		B: 1	2	3	$\bar{y}_{i..}$
A	1	32	16	18	22
	2	14	26	20	20
	$\bar{y}_{.j.}$	23	21	19	

We also have $a = 2$, $b = 3$, $r = 2$ and

$$\bar{y}_{..1} = 19.3333\ , \quad \bar{y}_{..2} = 22.6667\ , \quad \bar{y}_{...} = 21$$

Consequently, in each array and summing, we obtain

$$SST = \sum_{i=1}^{2}\sum_{j=1}^{3}\sum_{k=1}^{2}(y_{ijk} - \bar{y}_{...})^2$$

$$
\begin{aligned}
&= 8^2 + 14^2 + \cdots + (-5)^2 + 3^2 = 548 \\
SSA &= (2)(3)\sum_{i=1}^{2}(\bar{y}_{i..} - \bar{y}_{...})^2 = 6\left(1^2 + (-1)^2\right) = 12 \\
SSB &= (2)(2)\sum_{j=1}^{3}(\bar{y}_{.j.} - \bar{y}_{...})^2 = 4\left(2^2 + 0^2 + (-2)^2\right) = 32 \\
SSR &= (2)(3)\sum_{k=1}^{2}(\bar{y}_{..k} - \bar{y}_{...})^2 = 6\left((-1.6667)^2 + 1.6667^2\right) = 33.3347 \\
SS(AB) &= 2\sum_{i=1}^{2}\sum_{j=1}^{3}(\bar{y}_{ij.} - \bar{y}_{i..} - \bar{y}_{.j.} + \bar{y}_{...})^2 \\
&= 2\left(8^2 + (-6)^2 + (-2)^2 + (-8)^2 + 6^2 + 2^2\right) = 416 \\
SSE &= SST - SSA - SSB - SSR - SS(AB) = 54.6653
\end{aligned}
$$

The analysis-of-variance table is

Source of variation	Degrees of freedom	Sums of squares	Mean square	F
Replication	1	33.3347	33.3347	3.05
A	1	12.0000	12.0000	1.10
B	2	32.0000	16.0000	1.46
AB interaction:	2	416.0000	208.0000	19.03
Error	5	54.6653	10.9331	
Total	11	548.0000		

Since $F_{.05} = 6.61$ with 1 and 5 degrees of freedom, and $F_{.05} = 5.79$ with 2 and 5 degrees of freedom, only the AB interaction is significant at .05 level.

13.49 The visual summary of the four treatment means is given in Figure 13.4. According to the visual procedure, we assume that there is no replication effects in the model. The error sum of squares is

$$SSE = \sum_{i=0}^{1}\sum_{j=0}^{1}\sum_{l=0}^{r}(y_{ijl} - \bar{y}_{ij.})^2$$

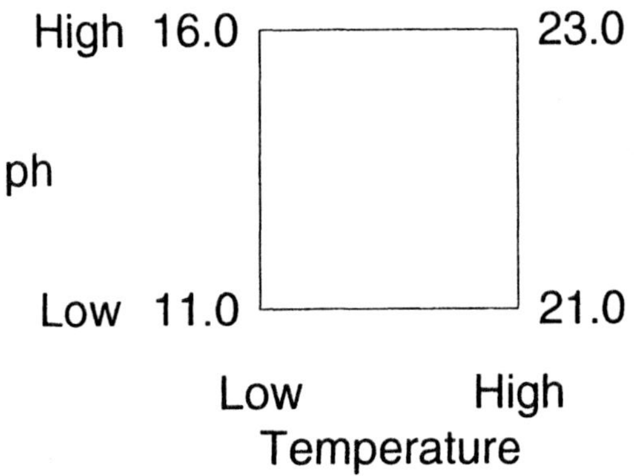

Figure 13.4: Visual summary of the experiment. Exercise 13.49

with $2^2(r-1)$ degrees of freedom. Thus we calculate

	Rep. 1	Rep. 2	Rep. 3	Mean	$y_{ij1}-\bar{y}_{ij.}$	$y_{ij2}-\bar{y}_{ij.}$	$y_{ij3}-\bar{y}_{ij.}$
1	10	14	9	11	−1	3	−2
a	21	19	23	21	0	−2	2
b	17	15	16	16	1	−1	0
ab	20	24	25	23	−3	1	2
Mean	17	18	18.25	17.75			

Squaring the entries in the last three columns and summing, we obtain

$$SSE = ((-1)^2 + 3^2 + (-2)^2 + \cdots + 2^2) = 38$$

with $2^2(r-1) = 4(2) = 8$ degrees of freedom. Therefore, $s^2 = MSE = 38/8 = 4.75$. Since $t_{.025} = 2.306$ for 8 degrees of freedom, the half length of the confidence intervals is

$$t_{.025}\sqrt{\frac{s^2}{r}} = 2.306\sqrt{\frac{4.75}{3}} = 2.90$$

and the confidence intervals are:

Factor A :	8.5 ± 2.9 or 5.6 to 11.4
Factor B :	3.5 ± 2.9 or .6 to 6.4
Interaction AB :	-1.5 ± 2.9 or -4.4 to 1.4

The confidence interval for interaction overs 0 but both those for main effects do not.

13.51 The visual summary of the eight treatment means is given in Figure 13.5. According to visual procedure, we assume that there is no replication effects in the model. The error sum of squares is

$$SSE = \sum_{i=0}^{1}\sum_{j=0}^{1}\sum_{k=0}^{1}\sum_{l=1}^{r}(y_{ijkl} - \bar{y}_{ijk.})^2$$

with $2^3(r-1)$ degrees of freedom. Thus we calculate

	Rep. 1	Rep. 2	Mean	$y_{ijk1} - \bar{y}_{ijk.}$	$y_{ijk2} - \bar{y}_{ijk.}$
1	41.8	42.2	42.0	$-.2$	.2
a	44.5	43.9	44.2	.3	$-.3$
b	56.5	56.3	56.4	.1	$-.1$
ab	57.3	56.5	56.9	.4	$-.4$
c	43.4	42.7	43.05	.35	$-.35$
ac	42.5	43.1	42.8	$-.3$	.3
bc	56.5	55.3	55.9	.6	$-.6$
abc	56.5	55.6	56.05	.45	$-.45$

Note that $y_{ijk1}-\bar{y}_{ijk.} = -(y_{ijk2}-\bar{y}_{ijk.})$. Squaring the entries in the last two columns and summing, or taking twice the sum of squares for one column, we obtain

$$SSE = 2(\,(-.2)^2 + (.3)^2 + \cdots + (.45)^2) = 2.150$$

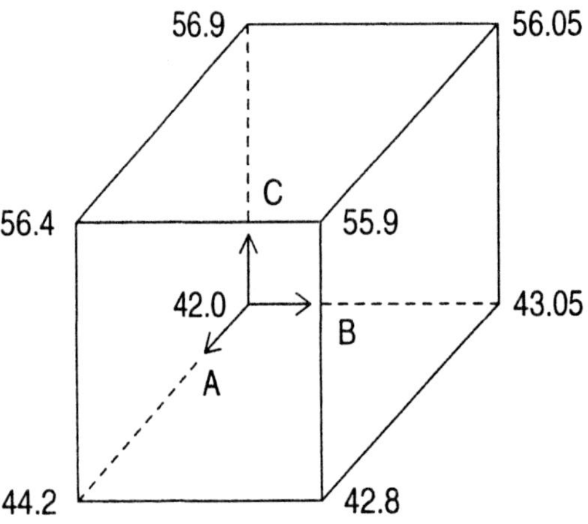

Figure 13.5: Visual summary of the experiment. Exercise 13.51

with $2^3(2-1) = 8$ degrees of freedom. Therefore, $s^2 = MSE = 2.150/8 = .26875$. Since $t_{.025} = 2.306$ for 8 degrees of freedom, the half length of the confidence intervals is

$$t_{.025}\sqrt{\frac{s^2}{2r}} = 2.306\sqrt{\frac{.26875}{4}} = .5977$$

The estimated effects are:

Factor A :

$$\bar{y}_{1...} - \bar{y}_{0...} = 49.9875 - 49.3375 = .6500$$

Factor B :

$$\bar{y}_{.1..} - \bar{y}_{.0..} = 56.3125 - 43.0125 = 13.3000$$

Factor C :

$$\bar{y}_{..1.} - \bar{y}_{..0.} = 49.4500 - 49.875 = -.4250$$

$AB:$

$$\frac{1}{2}(\bar{y}_{11..} - \bar{y}_{10..} - \bar{y}_{01..} + \bar{y}_{00..}) = \frac{1}{2}(56.4750 - 43.5000 - 56.1500 + 42.5250) = -.3250$$

$AC:$

$$\frac{1}{2}(\bar{y}_{1.1.} - \bar{y}_{1.0.} - \bar{y}_{0.1.} + \bar{y}_{0.0.}) = \frac{1}{2}(49.4250 - 50.5500 - 49.4750 + 49.2000) = -.7000$$

$BC:$

$$\frac{1}{2}(\bar{y}_{.11.} - \bar{y}_{.10.} - \bar{y}_{.01.} + \bar{y}_{.00.}) = \frac{1}{2}(55.9750 - 56.6500 - 42.9250 + 43.1000) = -.2500$$

$ABC:$

$$\frac{1}{4}(\bar{y}_{111.} - \bar{y}_{101.} - \bar{y}_{011.} + \bar{y}_{001.}) - \frac{1}{4}(\bar{y}_{110.} - \bar{y}_{100.} - \bar{y}_{010.} + \bar{y}_{000.})$$

$$= \frac{1}{4}(56.0500-42.8000-55.9000+43.0500) - \frac{1}{4}(56.9000-44.2000-56.4000+42.0000) = .5250$$

so the confidence intervals are:

$A:$	$6.5000 \pm .5977$ or 5.90 to 7.10
$B:$	$13.3000 \pm .5977$ or 12.70 to 13.90
$C:$	$-.4250 \pm .5977$ or -1.02 to $.017$
$AB:$	$.3250 \pm .5977$ or $-.27$ to $.92$
$AC:$	$-.7000 \pm .5977$ or -1.30 to $-.10$
$BC:$	$-.2500 \pm .5977$ or $-.85$ to $.35$
$ABC:$	$.5250 \pm .5977$ or $-.07$ to 1.12

The confidence interval for the AC interaction does not cover 0 so those two factors must be considered jointly. Factor B has the largest non-zero effect. Using the larger shot size increases tensile strength from 12.70 to 13.90 coded units.

13.53 Using the table in Exercise 13.49, we have

$$\begin{aligned} SST &= \sum_{i=0}^{1}\sum_{j=0}^{1}\sum_{l=1}^{r}(y_{ijl} - \bar{y}_{...})^2 \\ &= (-7.75)^2 + (-3.75)^2 + \cdots + 6.75^2 + 7.75^2 \;=\; 298.25 \\ SS(Tr) &= r\sum_{i=0}^{1}\sum_{j=0}^{1}(\bar{y}_{ij.} - \bar{y}_{...})^2 \\ &= 3((-6.75)^2 + (-1.75)^2 + (3.25)^2 + (5.25)^2) \;=\; 260.25 \end{aligned}$$

We use the Yates method to calculate the sum of squares of factors and interactions.

Yates method for Exercise 13.53

Exp. con.	Tr. total	1	2	Sum of squares
1	33	96	213	3,780.75
a	63	117	51	216.75
b	48	30	21	36.75
ab	69	21	−9	6.75

The analysis-of-variance table is

Source of variation	Degrees of freedom	Sums of squares	Mean square	F
A	1	216.75	216.75	45.63
B	1	36.75	36.75	7.74
AB	1	6.75	6.75	1.42
Error	8	38.00	4.75	
Total	11	298.25		

The critical values are $F_{.05} = 5.32$ and $F_{.01} = 11.26$, each with 1 and 8 degrees of freedom. The effect for A is significant at the .01 level. The effect for B is significant at the .05 level.

13.55 The largest number of blocks in which one can perform a 2^6 factorial experiment without confounding any main effects is $2^5 = 32$. It is clear that it is impossible to divide the experiment into 2^6 without confounding everything. If we confound on AB , BC , CD , DE, and EF we will have the desired blocks because the generalized interactions between these blocks do not include any main effects.

13.57 (a) The four blocks are

Block 1: a, abc, bd, cd, be, ce, ade, $abcde$
Block 2: b, c, ad, $abcd$, ae, $abce$, bde, cde
Block 3: ab, ac, d, bcd, e, bce, $abde$, $acde$
Block 4: 1, bc, abd, acd, abe, ace, de, $bcde$

(b) The block totals are(see Exercise 13.12):

Block 1:	Rep 1:	$4+2+3+10+11+4+15+16=65$
	Rep 2:	$9+4+7+6+5+8+9+11=59$
Block 2:	Rep 1:	$2+2+8+11+7+17+4+17=68$
	Rep 2:	$8+5+2+15+4+23+11+11=79$
Block 3:	Rep 1:	$15+11+0+6+3+4+10+5=54$
	Rep 2:	$7+6+3+14+7+8+6+10=61$
Block 4:	Rep 1:	$3+4+5+6+10+19+7+14=68$
	Rep 2:	$1+1+12+1+17+13+4+9=58$

Thus, $C = 512^2/64 = 4{,}096$ and the $SS(Bl) = 33{,}196/8 - 4{,}096 = 53.5$.

The analysis-of-variance table is

Source of variation	Degrees of freedom	Sums of squares 10^{-6}	Mean square 10^{-6}	F
Blocks	7	53.5000	7.6430	.5900
Main Effects:				
A	1	182.2500	182.2500	12.70
B	1	81.0000	81.0000	5.64
C	1	85.5625	85.5625	5.96
D	1	9.0000	9.0000	.63
E	1	248.0625	248.0625	17.28
Unconfounded Interactions:				
AB	1	16.0000	16.0000	1.17
AC	1	3.0625	3.0625	.21
AD	1	90.2500	90.2500	6.29
AE	1	7.5625	7.5625	.53
BC	1	7.5625	7.5625	.53
BD	1	1.0000	1.0000	.07
BE	1	.5625	.5625	.04
CD	1	22.5625	22.5625	1.57
CE	1	30.2500	30.2500	2.11
DE	1	10.5625	10.5625	.74
ABD	1	1.0000	1.0000	.07
ABE	1	3.0625	3.0625	.21
ACD	1	68.0625	68.0625	4.74
ACE	1	25.0000	25.0000	1.74
BCD	1	60.0625	60.0625	4.18
BCE	1	4.0000	4.0000	.28
BDE	1	60.0625	60.0625	4.18
CDE	1	36.0000	36.0000	2.51
$ABCD$	1	52.5625	52.5625	3.66
$ABCE$	1	9.0000	9.0000	.63
$ABDE$	1	18.0625	18.0625	1.26
$ACDE$	1	42.2500	42.2500	2.94
$ABCDE$	1	.2500	.2500	.02
Intrablock Error	28	401.8750	14.3527	
Total	63	1630.0000		

The critical value is $F_{.05} = 4.20$ with 1 and 28 degrees of freedom. The effects for A, B, C, E and AD, ACD interactions are significant at the .05 level. The result is similar to that of Exercise 13.12.

13.59 We choose the experimental conditions which have an even number of letters in common with the defining contrast $ABCDE$ and put them in the generalized standard order. The defining relation for the half replicate is $I = -ABCDE$. The table of calculations using the Yates method is given in the first table. We put "e" in parentheses to remind us that "e" has nothing to do with the analysis.

Yates method for Exercise 13.59

Exp. con.	Tr. total	1	2	3	4	Id.	Sum of squares
1	20	47	82	143	282	$I = -ABCDE$	4,970.25
$a(e)$	27	35	61	139	92	$A = -BCDE$	529.00
$b(e)$	9	38	76	43	−52	$B = -ACDE$	169.00
ab	26	23	63	49	26	$AB = -CDE$	42.25
$c(e)$	14	43	24	−27	−34	$C = -ABDE$	72.25
ac	24	33	19	−25	−4	$AC = -BDE$	1.00
bc	7	39	24	9	−8	$BC = -ADE$	4.00
$abc(e)$	16	24	25	17	−14	$ABC = -DE$	12.25
$d(e)$	18	7	−12	−21	−4	$D = -ABCE$	1.00
ad	25	17	−15	−13	6	$AD = -BCE$	2.25
bd	8	10	−10	−5	2	$BD = -ACE$	.25
$abd(e)$	25	9	−15	1	8	$ABD = -CE$	4.00
cd	15	7	10	−3	8	$CD = -ABE$	4.00
$acd(e)$	24	17	−1	−5	6	$ACD = -BE$	2.25
$bcd(e)$	4	9	10	−11	−2	$BCD = -AE$	.25
$abcd$	20	16	7	−3	8	$ABCD = -E$	4.00

We will assume that the two-way interactions are due to random error. The analysis-of-variance table is

Source of variation	Degrees of freedom	Sums of squares	Mean square	F
Confounded main effects:				
A	1	529.00	529.00	72.97
B	1	169.00	169.00	23.31
C	1	72.25	72.25	9.97
D	1	1.00	1.00	.14
E	1	4.00	4.00	.55
Error	10	72.50	7.25	
Total	15	847.75		

The critical value is $F_{.05} = 4.96$ with 1 and 10 degrees of freedom. The main effects for A, B and C are significant at the .05 level.

Chapter 14

THE STATISTICAL CONTENT OF QUALITY IMPROVEMENT PROGRAMS

14.1 (a) The center line for the $\bar{x}$ chart is given by $y = \mu = 0.150$. The lower control limit is given by

$$y = 0.150 - \frac{3}{\sqrt{5}}(0.002) = 0.147$$

and the upper control limit is given by

$$y = 0.150 + \frac{3}{\sqrt{5}}(0.002) = 0.153.$$

(b) The center line for the R chart is given by $y = d_2\sigma = 2.326(0.002) = 0.005$. The lower control limit is given by $y = D_1\sigma = 0(0.002) = 0$ and the upper control limit is given by $y = D_2\sigma = 4.918(0.002) = 0.010$.

(c) The control charts are given in the Figures 14.1 and 14.2. For the $\bar{x}$ chart, points at 8, 16, and 17 are outside the limits. For the R chart , all points are within the limits.

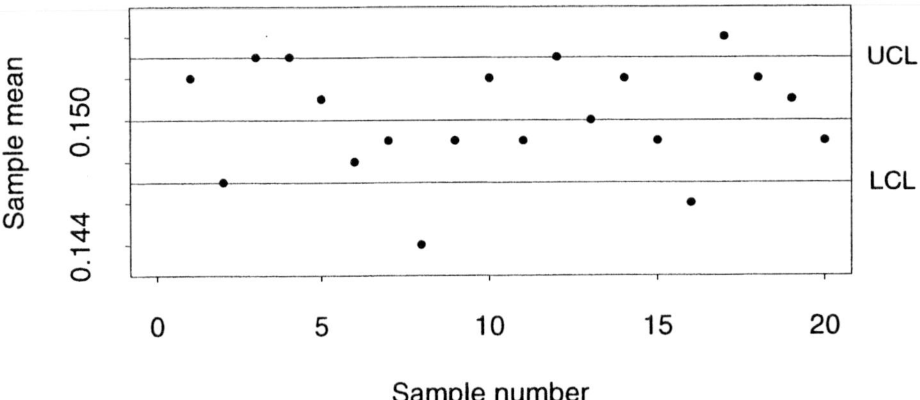

Figure 14.1: Control chart for sample means for Exercise 14.1.

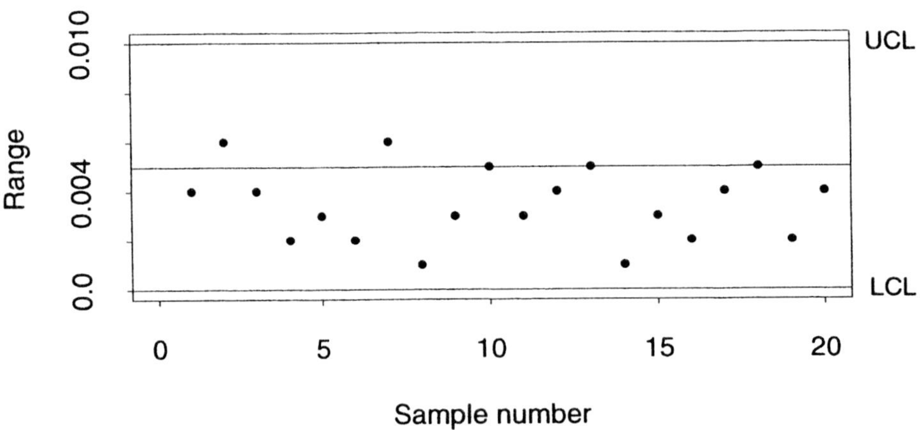

Figure 14.2: Control chart for sample ranges for Exercise 14.1.

14.3 We first find $\bar{\bar{x}} = 48.10$ and $\bar{R} = 2.95$.

(a) The center line for the $\bar{x}$ chart is $\bar{\bar{x}} = 48.10$, the $LCL = 45.95$ and the $UCL = 50.25$.

(b) The center line for the R chart is $\bar{R} = 2.95$, the $LCL = 0$ and the $UCL = 6.73$.

(c) The control charts are given in the Figures 14.3 and 14.4. For the $\bar{x}$ chart, there are points outside the limits at 5, 7, 8, 9, 11, 12, 13, 15, 16, 17, 18, 21, 24, and 25. For the R chart, all points are within the limits.

(d) Their are 8 runs with 11 points below and 13 above.

$$\mu_u = 12.92, \sigma_u = 2.38 \text{ so } z = -2.07$$

This is significant at the $\alpha = 0.025$ level ($z_\alpha = 1.96$).

(e) No, since the process in not in control.

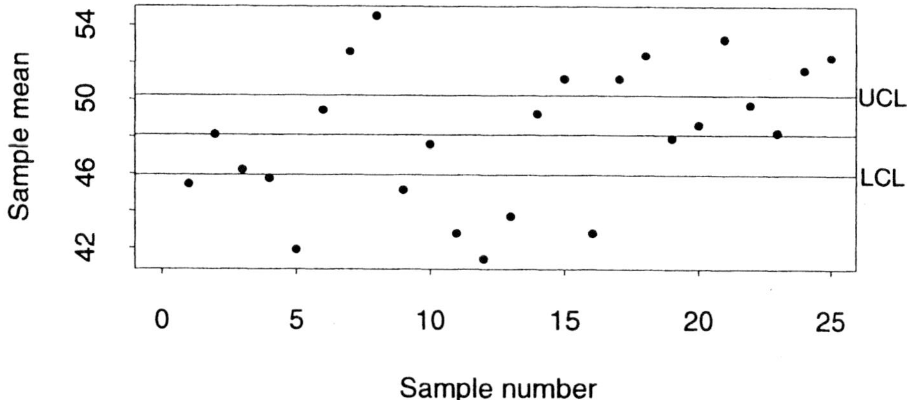

Figure 14.3: Control chart for sample means for Exercise 14.3.

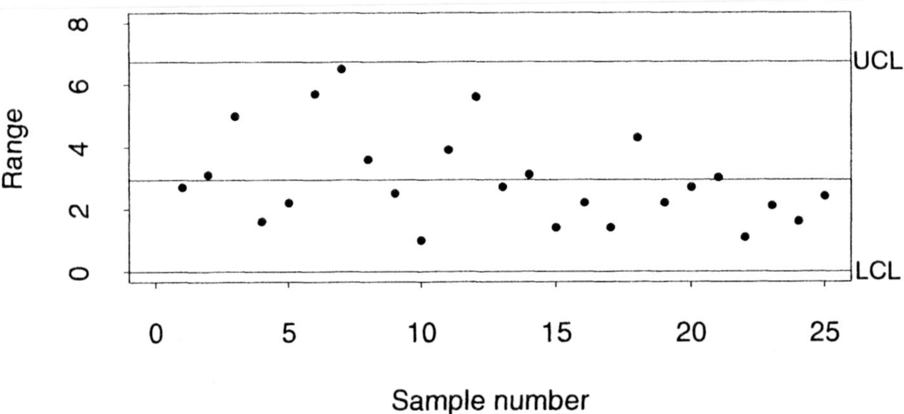

Figure 14.4: Control chart for sample ranges for Exercise 14.3.

14.5 (a) We calculate $\bar{\bar{x}} = 21.7$, $\bar{s} = 1.455$. Thus, for the $\bar{x}$ chart, the center line is 21.7 and

$$LCL = \bar{x} - A_1\bar{s} = 21.7 - 2.394(1.455) = 18.22$$

$$UCL = \bar{x} + A_1\bar{s} = 21.7 + 2.394(1.455) = 25.18$$

The σ chart has center line $c_2\bar{s} = .72356\ (\ 1.455) = 1.053$

$$LCL = B_3\bar{s} = 0(1.445) = 0$$

$$UCL = B_4\bar{s} = 2.568(1.445) = 3.736$$

The control charts are given in Figures 14.5 and 14.6. All of the sample means are within the control limits. Only one sample standard deviation is outside the control limits, namely the one for the $17th$ sample.

(b) Yes. The process is in control.

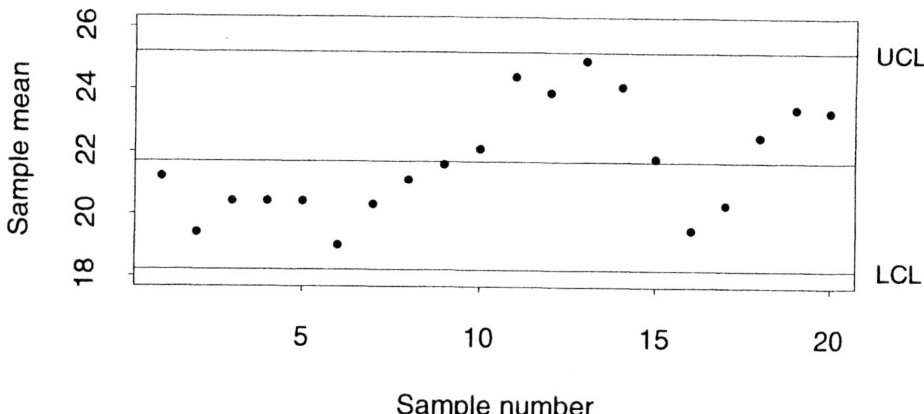

Figure 14.5: Control chart for sample means for Exercise 14.5.

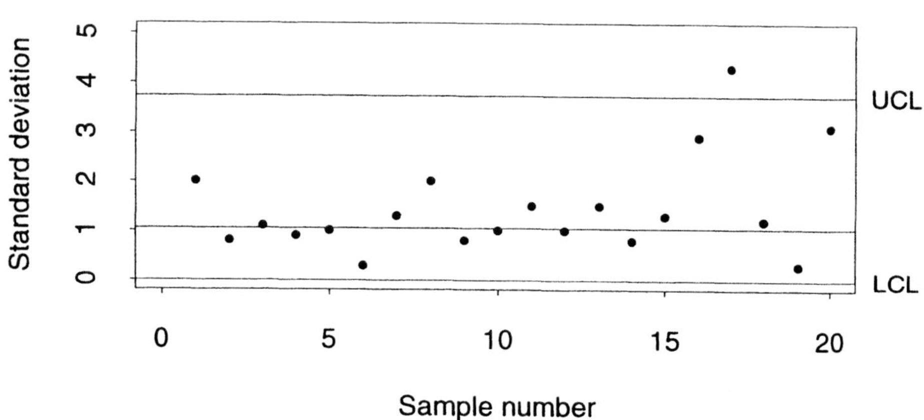

Figure 14.6: Control chart for sample ranges for Exercise 14.5.

14.7 The σ chart has center line $c_2\bar{s} = 0.8407\ (\ 0.002) = 0.00168$.

$$LCL = B_3\bar{s} = 0(0.002) = 0$$

$$UCL = B_4\bar{s} = 2.089(0.002) = 0.00418.$$

14.9 (a) For the data of Exercise 14.8, $\bar{p} = 0.0369$. Thus,

$$\begin{aligned} LCL &= \bar{p} - 3\sqrt{\bar{p}(1-\bar{p})/100} \\ &= 0.0369 - 3\sqrt{0.0369(1-0.0369)/100} = -0.020 \end{aligned}$$

which is taken to be 0, and

$$\begin{aligned} UCL &= \bar{p} + 3\sqrt{\bar{p}(1-\bar{p})/100} \\ &= 0.0369 + 3\sqrt{0.0369(1-0.0369)/100} = 0.0935 \end{aligned}$$

(b) The control chart is given in Figure 14.7. The process is in control.

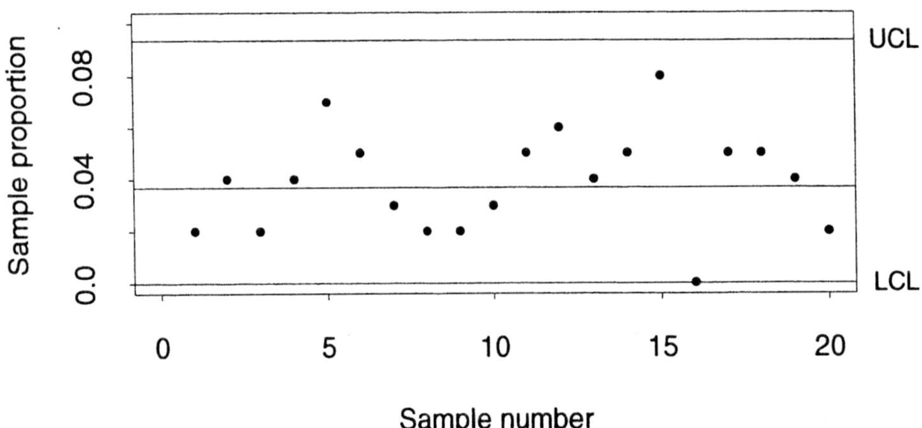

Figure 14.7: Control chart for the proportion defective for Exercise 14.9.

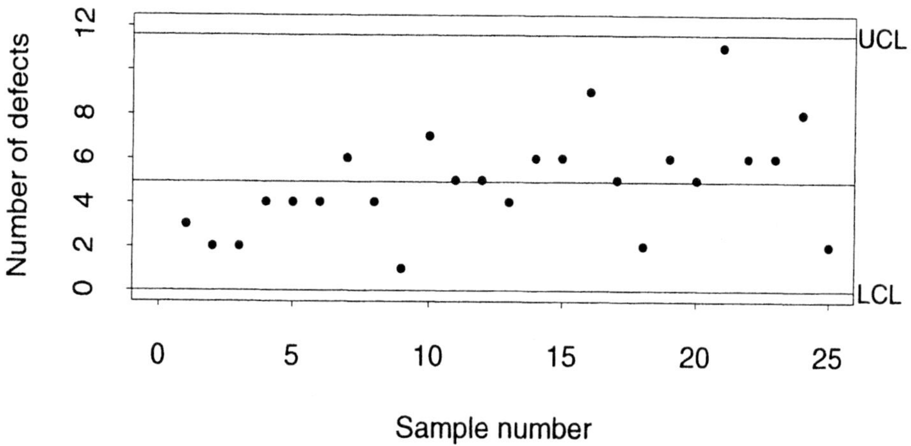

Figure 14.8: Control chart for the proportion defective for Exercise 14.11.

14.11 The center line is $\bar{c} = 4.92$. The LCL is

$$LCL = \bar{c} - 3\sqrt{\bar{c}} \;=\; -0.173$$

so the LCL is taken to be 0. The UCL is

$$UCL = \bar{c} + 3\sqrt{\bar{c}} \;=\; 11.57$$

The control chart is given in Figure 14.8. The process is in control.

14.13 From Table 14, the tolerance limits are

$$\bar{x} \pm Ks = 52,800 \pm 3.457(4600)$$

or, from 36,897.8 to 68,702.2 . This means that with 95 percent confidence, 99 percent of the pieces will have a yield stress between 36,897.8 and 68,702.2 psi.

14.15 (a) $\bar{x}$ =0.1063, $s = 0.0040$, $K = 2.052$. Thus, the tolerance limits are

$$\bar{x} \pm Ks = 0.1063 \pm 2.052(0.0004)$$

or, from 0.1054 to 0.1072 .

(b) The confidence interval is given by

$$\bar{x} - t_{\alpha/2}\frac{s}{\sqrt{n}} < \mu < \bar{x} + t_{\alpha/2}\frac{s}{\sqrt{n}}$$

Since $n = 40$, the student's t quantile $t_{\alpha/2} = t_{.025}$ is well approximated by the standard normal quantile $z_{.025} = 1.96$. The 95 percent confidence interval is

$$.1063 - 1.96\frac{.0004}{\sqrt{40}} < \mu < .1063 + 1.96\frac{.0004}{\sqrt{40}}$$

or $.1062 < \mu < .1064$. For the difference between tolerance limits and confidence intervals, see the discussion in Section 14.4 of the text.

14.17 (a) Since the lot size is very large, we can use the binomial distribution. Thus, we need the probability of three or fewer for a binomial distribution with $p =$ .15 and $n = 50$ to determine the probability of accepting a lot when the true proportion defective is .15. This is given by

$$\binom{50}{0}(.15)^0(.85)^{50} + \binom{50}{1}(.15)^1(.85)^{49}$$

$$+\binom{50}{2}(.15)^2(.85)^{48} + \binom{50}{3}(.15)^3(.85)^{47}$$

$$= 2.9576 \times 10^{-4} + 2.6097 \times 10^{-3} + 0.011283 + 0.031858 = 0.046047.$$

If the true proportion of defectives is .04, the probability of rejecting is

$$1-\binom{50}{0}(.04)^0(.96)^{50}-\binom{50}{1}(.04)^1(.96)^{49}-\binom{50}{2}(.04)^2(.96)^{48}$$

$$-\binom{50}{3}(.04)^3(.96)^{47}=1-.86087=.13913.$$

(b) Using the Poisson approximation, when the true proportion is .15, the expected number of defectives in 50 is $50(.15) = 7.5$. Interpolating in Table 2, we see that the probability of 3 or less defectives is .059. The expected number of defectives when the true proportion is .04 is $50(.04) = 2$. From Table 2, the probability of three or fewer is .857, so the probability of more than 3 is $1 - .8574 = .143$.

14.19 The calculations are given in Table 4.1 where $\lambda = 50p$ and

$$L(p) = e^{-\lambda}(1+\lambda+\frac{\lambda^2}{2}+\frac{\lambda^3}{6})$$

Table 14.1. OC curve for Exercise 14.19.

p	λ	$L(p)$	p	λ	$L(p)$
.01	0.5	.9982	.11	5.5	.2017
.02	1.0	.9810	.12	6.0	.1512
.03	1.5	.9344	.13	6.5	.1118
.04	2.0	.8571	.14	7.0	.0818
.05	2.5	.7576	.15	7.5	.0591
.06	3.0	.6472	.16	8.0	.0424
.07	3.5	.5366	.17	8.5	.0301
.08	4.0	.4335	.18	9.0	.0212
.09	4.5	.3423	.19	9.5	.0149
.10	5.0	.2650	.20	10.0	.0103

The *OC* curve is given in Figure 14.9. The producers risk is $1 - .86 = .14$ and consumers risk is .08 .

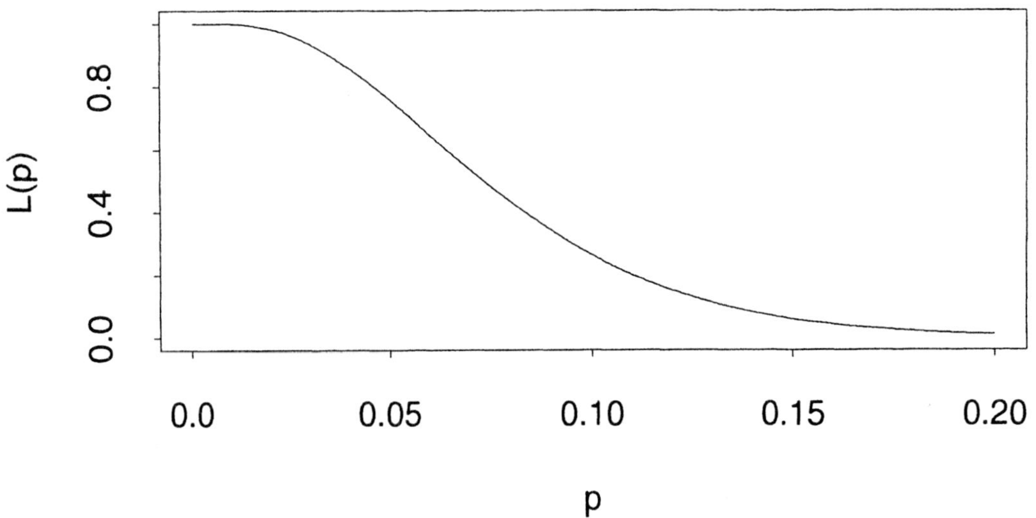

Figure 14.9: OC curve for Exercise 14.19.

14.21 (a) Using the hypergeometric distribution

$$L(.10) = \frac{\binom{10}{0}\binom{90}{10} + \binom{10}{1}\binom{90}{9}}{\binom{100}{10}}$$

$$= \frac{(90!/80!10!) + 10(90!/9!81!)}{100!/90!10!} = .330476 + .407995 = .73847$$

(b)

$$L(.10) = \binom{10}{0}(.10)^0(.90)^10 + \binom{10}{1}(.10)^1(.90)^9$$

$$= (.90)^9(.90 + 10(.10)) = .73610.$$

(c)

$$L(p) = (1-p)^9(1-p+10p) = (1-p)^9(1+9p)$$

The calculations for the plot of the OC curve are in Table 14.2. The plot of the OC curve is given in Figure 14.10.

(d) $AOQ = pL(p)$. The calculations for the plot are also shown in Table 14.2 and the plot is given in Figure 14.11.

Table 14.2. Calculation of the OC curve and AOQ curve for Exercise 14.21.

p	$L(p)$	AOQ	p	$L(p)$	AOQ
0.03	0.9655	0.0290	0.33	0.1080	0.0356
0.06	0.8824	0.0529	0.36	0.0764	0.0275
0.09	0.7746	0.0697	0.39	0.0527	0.0206
0.12	0.6583	0.0790	0.42	0.0355	0.0149
0.15	0.5443	0.0816	0.45	0.0233	0.0105
0.18	0.4392	0.0791	0.48	0.0148	0.0071
0.21	0.3464	0.0727	0.51	0.0091	0.0046
0.24	0.2673	0.0642	0.54	0.0054	0.0029
0.27	0.2019	0.0545	0.57	0.0031	0.0018
0.30	0.1493	0.0448	0.60	0.0017	0.0010

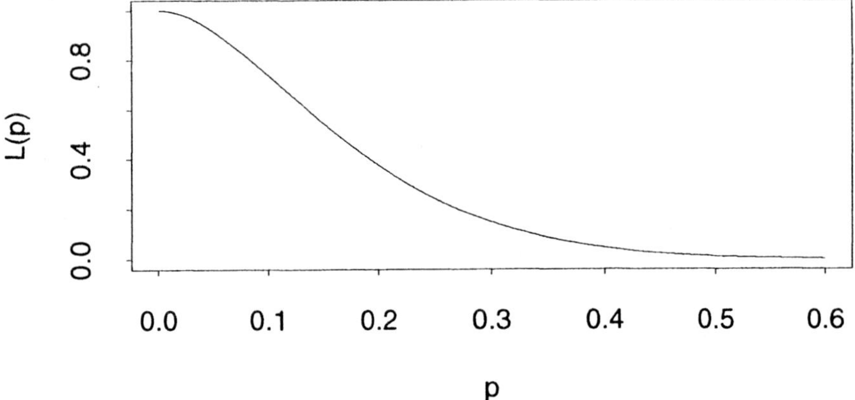

Figure 14.10: OC curve for Exercise 14.21.

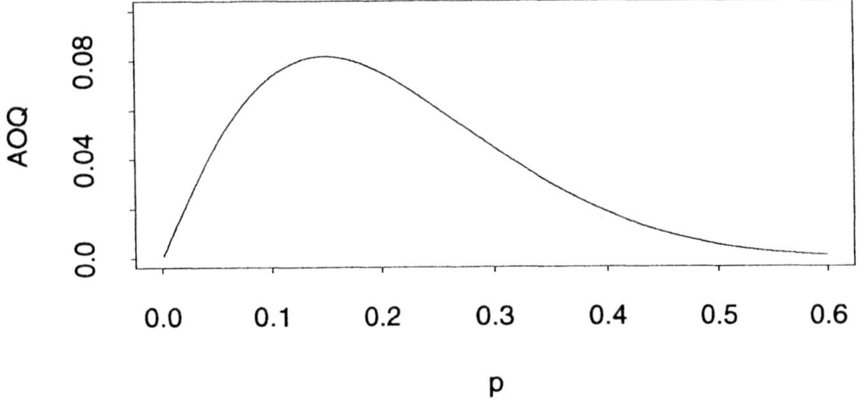

Figure 14.11: AOQ curve for Exercise 14.21.

14.23 The AOQ curve is given by $p \cdot L(p)$. The calculations are given in Table 14.3 and the AOQ curve in Figure 14.12. From that figure, we see that the maximum occurs at $p = .0225$ with a maximum AOQ of .01550. Thus, the $AOQL = .0155$.

Table 14.3. Calculations for the AOQ curve for Exercise 14.23.

p	AOQ	p	AOQ	p	AOQ
0.0025	0.0025	0.0275	0.0150	0.0525	0.0066
0.0050	0.0050	0.0300	0.0144	0.0550	0.0059
0.0075	0.0075	0.0325	0.0136	0.0575	0.0052
0.0100	0.0098	0.0350	0.0127	0.0600	0.0046
0.0125	0.0119	0.0375	0.0118	0.0625	0.0041
0.0150	0.0135	0.0400	0.0109	0.0650	0.0036
0.0175	0.0146	0.0425	0.0100	0.0675	0.0032
0.0200	0.0153	0.0450	0.0091	0.0700	0.0028
0.0225	0.0155	0.0475	0.0082	0.0725	0.0024
0.0250	0.0154	0.0500	0.0074	0.0750	0.0021

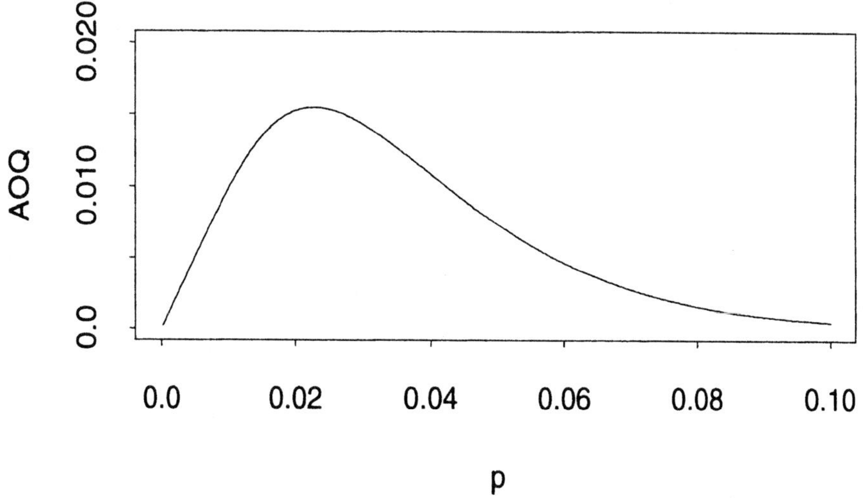

Figure 14.12: OC curve for Exercise 14.23.

14.25 The producer's risk is

$$P(5 \text{ or more in the 1-st sample})$$

$$+P(1 \text{ in the 1-st sample}) \cdot P(5 \text{ or more in the 2-nd sample})$$

$$+P(2 \text{ in the 1-st sample}) \cdot P(4 \text{ or more in the 2-nd sample})$$

$$+P(3 \text{ in the 1-st sample}) \cdot P(3 \text{ or more in the 2-nd sample})$$

$$+P(4 \text{ in the 1-st sample}) \cdot P(2 \text{ or more in the 2-nd sample})$$

or

$$(1 - .9873) + (.3431)(.0641) + (.2669)(.1807) + (.1285)(.3805)$$

$$+ \ (.0429)(.6195) = .1584.$$

14.27 The code letter is G. The sampling plan is to randomly select 32 items, accept if there are 5 or fewer defectives, and reject if there are 6 or more.

14.29 Using the formulas in Section 14.9 of the book,

$$a_n = -1.67 + (.186)n$$

$$r_n = 2.14 + (.186)n$$

The acceptance and rejection numbers for 10 trials are given in Table 14.4. The sample would be rejected on the 8-th trial.

Table 14.4. Acceptance and rejection numbers. Exercise 4.29.

Trial no.	Acceptance number	Rejection number
1	-	-
2	-	-
3	-	3
4	-	3
5	-	4
6	-	4
7	-	4
8	-	4
9	0	4
10	0	5

14.31 (a) The center line for the $\bar{x}$ chart is given by $y = \mu = 4.1$. The lower control limit is given by

$$y = 4.1 - \frac{3}{\sqrt{5}}(0.05) = 4.013$$

and the upper control limit is given by

$$y = 4.1 + \frac{3}{\sqrt{5}}(0.05) = 4.187.$$

(b) The center line for the R chart is given by $y = d_2\sigma = 2.326(0.05) = 0.163$. The lower control limit is given by $y = D_1\sigma = 0(0.05) = 0$ and the upper control limit is given by $y = D_2\sigma = 4.918(0.05) = 0.246$.

(c) The control charts are given in the Figures 14.13 and 14.14. For the $\bar{x}$ chart, points at 1, 3, 4, 5, 7, 8, 9, and 11 to 20 are outside the limits. For the R chart, the point at 6 is the only one outside of the limits.

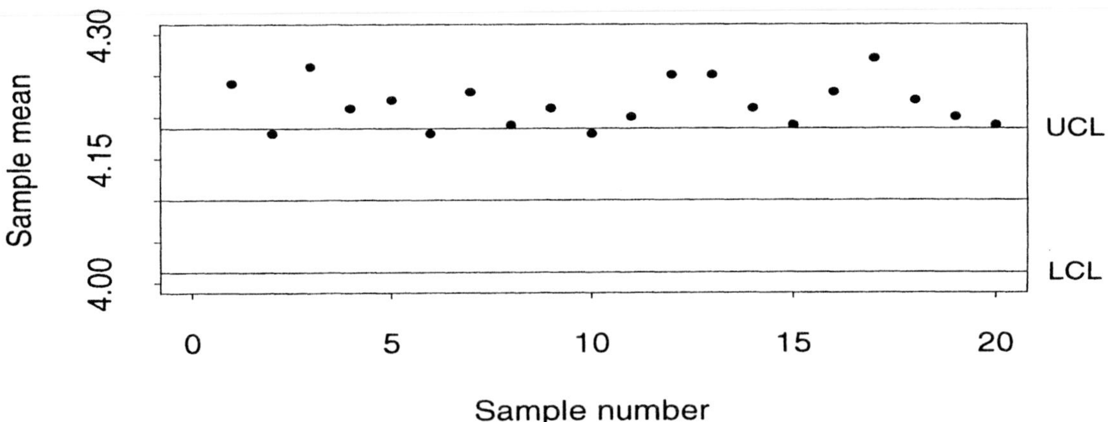

Figure 14.13: Control chart for the sample means for Exercise 14.31.

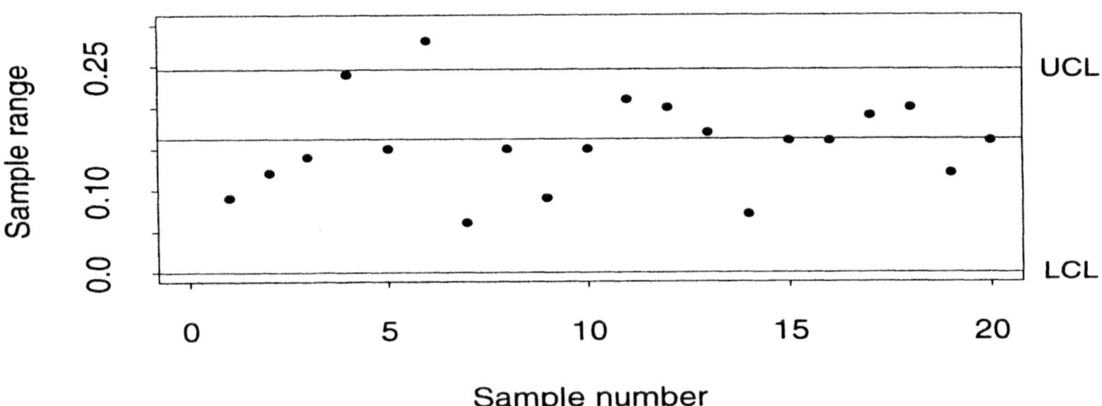

Figure 14.14: Control chart for the fraction defective for Exercise 14.33.

14.35 The center line is $\bar{c} = 0.8$. The LCL is

$$LCL = \bar{c} - 3\sqrt{\bar{c}} \;=\; -1.88$$

so the LCL is taken to be 0. The UCL is

$$UCL = \bar{c} + 3\sqrt{\bar{c}} \;=\; 3.48$$

The control chart is given in Figure 14.15. Only the point at 19 is outside of the limits.

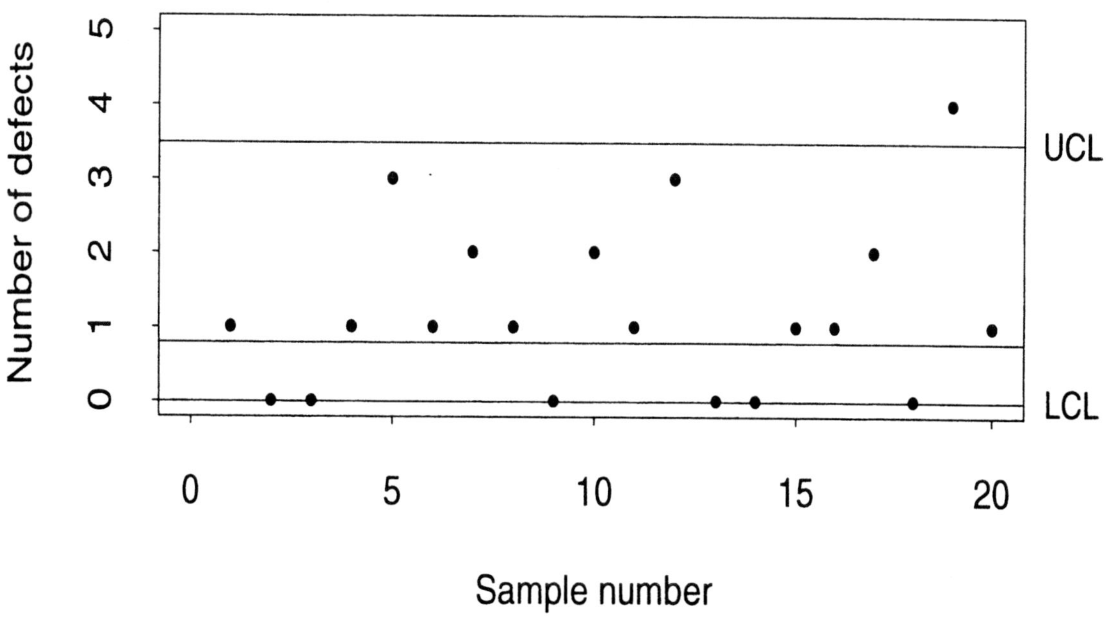

Figure 14.15: Control chart for number of defects for Exercise 14.35.

14.37 The $n = 50$ transformed observations $y = \ln x$ have $\bar{y} \;=\; 8.846$ and $s_y = 1.0293$

From Table 14 $K = 1.996$ so the 95 percent tolerance limits on proportion $P = .90$ for the transformed observations are

$$\bar{x} \pm Ks = 8.846 \pm 1.996(1.2093)$$

or 6.7915 to 10.9005. Converting these to the original scale, the tolerance limits are exp(6.7915) to exp(10.9005) or 890 to 54,203. This means that, with 95 percent confidence, 90 percent of the inter-request times will be between 890 and 54,203.

14.39 The calculations are given in Table 14.5 where $\lambda = 40p$ and

$$L(p) = e^{-\lambda}(1 + \lambda + \frac{\lambda^2}{2})$$

Table 14.5. Calculation of OC curve for Exercise 14.39.

p	λ	$L(p)$	p	λ	$L(p)$
.01	0.4	.9921	.11	4.4	.1851
.02	0.8	.9526	.12	4.8	.1425
.03	1.2	.8795	.13	5.2	.1088
.04	1.6	.7834	.14	5.6	.0824
.05	2.0	.6767	.15	6.0	.0620
.06	2.4	.5697	.16	6.4	.0463
.07	2.8	.4695	.17	6.8	.0344
.08	3.2	.3799	.18	7.2	.0255
.09	3.6	.3027	.19	7.6	.0188
.10	4.0	.2381	.20	8.0	.0138

The *OC* curve is presented in Figure 14.16. The producer's risk is $1 - .78 = .22$ and the consumer's risk is .08.

14.41 The code letter is F. The sampling plan is to randomly select 20 items, accept if there are 2 or fewer defectives, and reject if there are 3 or more.

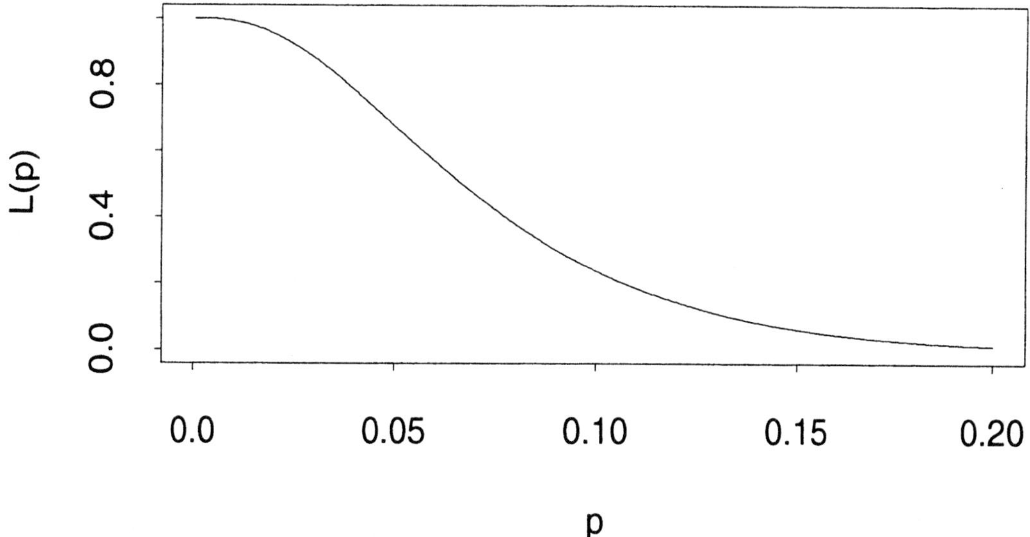

Figure 14.16: OC curve for Exercise 14.39.

14.43 (a) We check that $\bar{x}$ =237.32, $s = 25.10$, $K = 2.126$. Thus, the tolerance bound is

$$\bar{x} - Ks = 237.32 - 2.126(25.10)$$

or 183.

(b) Here $K = 2.010$. Thus, the tolerance bound is

$$\bar{x} - Ks = 237.32 - 2.010(25.10)$$

or 186.

(c) The normal-scores plot, given in Figure 14.17, reveals no marked departures from the assumption of normal observations.

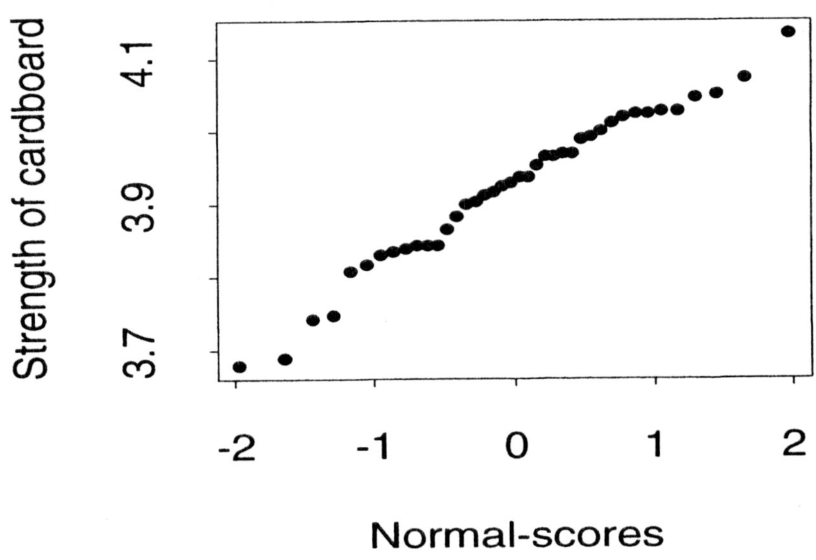

Figure 14.17: Normal scores plot of cardboard data. Exercise 14.43.

14.45 The sample mean and sample standard deviation are

$$\bar{x} = 2.5064 \quad \text{and} \quad s = .0022.$$

(a)

$$\hat{C}_p = \frac{USL - LSL}{6s} = \frac{2.516 - 2.496}{6(.0022)} = 1.515.$$

(b)

$$\begin{aligned}\hat{C}_{pk} &= \frac{\min(\bar{x} - LSL,\ USL - \bar{x})}{3s} \\ &= \frac{\min(2.5064 - 2.4960,\ 2.5160 - 2.5064)}{3(.0022)} \\ &= \frac{\min(.0104,\ .0096)}{3(.0022)} = 1.45\end{aligned}$$

14.47 The sample mean and sample standard deviation are

$$\bar{x} = -3.2432 \quad \text{and} \quad s = 10.2886.$$

(a)

$$\hat{C}_p = \frac{USL - LSL}{6s} = \frac{50 - (-50)}{6(10.2886)} = 1.62.$$

(b)

$$\begin{aligned}
\hat{C}_{pk} &= \frac{\min(\bar{x} - LSL,\ USL - \bar{x})}{3s} \\
&= \frac{\min(-3.2432 - (-50),\ 50 - (-3.2432))}{3(10.2886)} \\
&= \frac{\min(46.7568,\ 53.2432)}{3(10.2886)} = 1.515.
\end{aligned}$$

Chapter 15

APPLICATIONS TO RELIABILITY AND LIFE TESTING

15.1 Since this is a series system, we need to find R such that

$$R^8 = .95 \quad \text{so} \quad R = (.95)^{1/8} = .9936.$$

15.3 The reliability of the B, C parallel segment is :

$$R_{BC} = 1 - (1 - .80)(1 - .90) = .980.$$

The reliability of the E, F parallel segment is :

$$R_{EF} = 1 - (1 - .90)(1 - .85) = .985.$$

The reliability of the A, B, C combination segment is :

$$R_{ABC} = R_A\, R_{BC} = (.95)(.980) = .931.$$

The reliability of the D, E, F combination segment is :

$$R_{DEF} = R_D\, R_{EF} = (.99)(.985) = .97515.$$

Thus, the reliability of the system is:

$$\begin{aligned} R &= 1 - (1 - R_{ABC})(1 - R_{DEF}) \\ &= 1 - (1 - .931)(1 - .97515) = .9983. \end{aligned}$$

15.5 (a) We have that

$$f(t) = Z(t)\, exp\,[\, -\int_0^t Z(t)\; dt\,].$$

where

$$Z(t) = \begin{cases} \beta(1 - t/\alpha) & \text{for } 0 < t < \alpha. \\ 0 & \text{elsewhere.} \end{cases}$$

Thus

$$f(t) = \begin{cases} \beta(1 - t/\alpha)\, exp\,[\, -\int_0^t \beta(1 - x/\alpha)\; dx\,] & \text{for } 0 < t < \alpha. \\ 0 & \text{elsewhere.} \end{cases}$$

so

$$f(t) = \begin{cases} \beta(1 - t/\alpha)\, exp\,[\, -\beta(t - t^2/(2\alpha)\,)\,] & \text{for } 0 < t < \alpha. \\ 0 & \text{elsewhere.} \end{cases}$$

The distribution function is

$$F(t) = \begin{cases} \int_0^t f(x)\; dx & \text{for } 0 < t < \alpha. \\ \int_0^\alpha f(x)\; dx & \text{for } t > \alpha. \end{cases}$$

or

$$F(t) = \begin{cases} 1 - exp\,[\,-\beta(t - t^2/(2\alpha)\,)\,] & \text{for } 0 < t < \alpha. \\ 1 - exp\,[\,-\alpha\beta/2] & \text{for } t > \alpha. \end{cases}$$

(b) the probability of initial failure is $F(\alpha) = 1 - e^{-\alpha\beta/2}$ as shown in the previous part.

15.7 (a) Since the failure rate is constant,

$$f(t) = (.02)e^{-(.02)t}$$

and

$$F(t) = 1 - e^{-(.02)t}.$$

The unit of time is 1000 hours. The probability that the chip will last longer than 20,000 hours is

$$1 - F(20) = e^{-(.02)\cdot 20} = .6703.$$

(b) The 5000-hour reliability of four such chips in a series is

$$R = [1 - F(5)]^4 = e^{-4(.02)\cdot 5} = .6703.$$

15.9 Using the formula for the mean time between failures

$$\mu_p = \frac{1}{\alpha}(1 + \frac{1}{2} + + \cdots + \frac{1}{n}) = \frac{1}{9 \times 10^{-4}}(1 + \frac{1}{2} + + \cdots + \frac{1}{n}).$$

Since the failure rate is the inverse of the mean time between failures, in this case we require that $\mu_p \geq 1/(4 \times 10^{-4})$. Thus we must find n such that

$$\frac{9}{4} \leq (1 + \frac{1}{2} + \cdots + \frac{1}{n}).$$

A trial and error search yields $n = 5$ and the failure rate is

$$\frac{9.0 \times 10^{-4}}{2.2833} = 3.94 \times 10^{-4}.$$

15.11 (a) The probability that a part will last 500 hours is:

$$1 - F(500) = e^{-500/1000} = .6065.$$

(b) The probability that at least one will fail is one minus the probability that none will fail or

$$1 - e^{-3(1000)/1000} = .9502.$$

(c) The probability that a particular part fails is:

$$1 - e^{-600/1000} = .4512.$$

The probability that exactly two parts fail is given by the binomial distribution with $p = .4512$ or

$$\binom{4}{2} (.4512)^2 \, (1 - .4512)^2 = .3679.$$

15.13 (a) The probability of failure on any trial is p and the probability of no failure is $1 - p$ provided failure has not occurred previously. thus, failure on the x'th trial means no failure for $x - 1$ trials and then a failure. Since the trials are independent

$$f(x) = (1-p)^{x-1}\, p \quad \text{for} \quad x = 1, 2, 3, \ldots$$

(b)

$$F(x) \; = \; \sum_{i=1}^{x} (1-p)^{i-1} p = p \sum_{i=0}^{x-1} (1-p)^i$$

$$= p\frac{1-(1-p)^x}{p} = 1-(1-p)^x \quad \text{for} \quad x = 1,2,3,\cdots$$

(c) The probability that the switch survives 2,000 cycles is

$$1 - F(2000) = (1 - 6 \times 10^{-4})^{2000} = .301.$$

15.15 (a)

$$\begin{aligned} T_r &= \sum_{i=1}^{r} t_i + (n-r)t_r \\ &= 250 + 380 + 610 + 980 + 1250 + 30 \cdot 1250 = 40,970. \end{aligned}$$

Since $\chi^2_{.995}$ with 10 degrees of freedom is 2.156 and $\chi^2_{.005}$ with 10 degrees of freedom is 25.118, the 99 percent confidence interval is

$$\frac{2 \cdot 40,970}{25.118} < \mu < \frac{2 \cdot 40,970}{2.156}$$

or

$$3,253.1 < \mu < 38,005.6$$

(b) The null hypothesis is $\mu = 5,000$ and the alternative is $\mu > 5,000$. Since $\chi^2_{.05}$ with $2 \cdot r = 10$ degrees of freedom is 18.307, we reject the null hypothesis at the .05 level if

$$T_r > \frac{1}{2}\mu_0\chi^2_{.05} = \frac{1}{2}(5,000)(18.307) = 45,767.5$$

Since $T_r =$ 40,970 , we cannot reject the null hypothesis at the .05 level of significance. We cannot be sure the manufacturer's claim is true.

15.17 (a)

$$\begin{aligned} T_r &= \sum_{i=1}^{r} t_i + (n-r)t_r \\ &= 211 + 350 + 384 + 510 + 539 + 620 + 715 = 3,329. \end{aligned}$$

Since $\chi^2_{.975}$ with 14 degrees of freedom is 5.629 and $\chi^2_{.025}$ with 14 degrees of freedom is 26.119, the 95 percent confidence interval is

$$\frac{2 \cdot 3329}{26.119} < \mu < \frac{2 \cdot 3329}{5.629}$$

or

$$254.9 < \mu < 1,182.8$$

(b) The null hypothesis is $\mu = 500$ and the alternative is $\mu \neq 500$. Since $\chi^2_{.05}$ with 14 degrees of freedom is 23.658 and $\chi^2_{.95}$ with 14 degrees of freedom is 6.571, we reject the null hypothesis at the .10 level if

$$T_r < \frac{1}{2}\mu_0\chi^2_{.95} = \frac{1}{2}(500)(6.571) = 1,642.75$$

or if

$$T_r > \frac{1}{2}\mu_0\chi^2_{.05} = \frac{1}{2}(500)(23.685) = 5,921.25$$

Since T_r = 3,329 , we cannot reject the null hypothesis at the .10 level of significance.

15.19 Since $2T_r/\mu$ is a χ^2 random variable with $2r$ degrees of freedom ,

$$P[\ \chi^2_{1-\alpha/2} < \frac{2T_r}{\mu} \ \ and \ \ \frac{2T_r}{\mu} < \chi^2_{\alpha/2}\ ] = \alpha,$$

where $\chi^2_{1-\alpha/2}$ and $\chi^2_{\alpha/2}$ are the chi-square quantiles for $2r$ degrees of freedom.

Multiplying the first inequality by $\mu/\chi^2_{1-\alpha/2}$ and the second by $\mu/\chi^2_{\alpha/2}$ gives:

$$P\left[\ \mu < \frac{2T_r}{\chi^2_{1-\alpha/2}} \text{ and } \frac{2T_r}{\chi^2_{\alpha/2}} < \mu\ \right] = \alpha.$$

Thus the $(1-\alpha)100$ percent confidence interval is:

$$\frac{2T_r}{\chi^2_{\alpha/2}} < \mu < \frac{2T_r}{\chi^2_{1-\alpha/2}}$$

15.21 There are $r = 3$ failures and total time on test is

$$T_3 = 2076 + 3667 + 9102 + 197(9102) = 1,807,939$$

Since $\chi^2_{.05} = 12.592$ for $2r = 6$ degrees of freedom, the 95 percent lower confidence bound is

$$\frac{2T_3}{\chi^2_{.05}} = \frac{2(1,807,939)}{12.592} = 287,156.8$$

15.23 The probability the diaphragm valve will perform at least 150 hours is 1 minus the probability that it fails before 150 hours. Since

$$1 - F(150) = 1 - (1 - e^{-\alpha(150)^{\beta}}),$$

$\hat{\alpha} = .0105$ and $\hat{\beta} = .5062$, we estimate this probability by

$$e^{-\hat{\alpha}(150)^{\hat{\beta}}} = e^{-(.0105)(150)^{.5062}} = .8758.$$

15.25 Suppose T has the Weibull distribution with parameters α and β. In the text it was shown that

$$E(T) = \frac{1}{\alpha^{1/\beta}}\Gamma(1 + \frac{1}{\beta})$$

Since $Var(T) = E(T^2) - [E(T)]^2$, we must find $E(T^2)$. The Weibull density is

$$\alpha\beta t^{\beta-1}e^{-\alpha t^\beta}$$

so

$$E(T^2) = \int_0^\infty t^2\alpha\beta t^{\beta-1}e^{-\alpha t^\beta}\,dt.$$

Let $u = \alpha t^\beta$. Then

$$E(T^2) = \frac{1}{\alpha^{2/\beta}}\int_0^\infty u^{2/\beta}e^{-u}du = \frac{1}{\alpha^{2/\beta}}\Gamma(1+\frac{2}{\beta})$$

so that

$$\begin{aligned} Var(T) &= \frac{1}{\alpha^{2/\beta}}\Gamma(1+\frac{2}{\beta}) - (\frac{1}{\alpha^{1/\beta}}\Gamma(1+\frac{1}{\beta}))^2 \\ &= \frac{1}{\alpha^{2/\beta}}\cdot[\,\Gamma(1+\frac{2}{\beta}) - (\Gamma(1+\frac{1}{\beta}))^2\,] \end{aligned}$$

15.27 (a) The probability of failure during the first 250 hours of operation is:

$$1 - e^{-(.0045)\cdot 250} = .6753.$$

(b) The probability that two independent components will survive the first 100 hours of operation is:

$$e^{-(.0045)\cdot 100}\cdot e^{-(.0045)\cdot 100} = .4066.$$

15.29 The mean time between failures for a series system is:

$$MTBF = \frac{1}{\frac{1}{\mu_1}+\frac{1}{\mu_2}+\frac{1}{\mu_3}+\frac{1}{\mu_4}+\frac{1}{\mu_5}+\frac{1}{\mu_6}}$$

$$= \frac{1}{1.8 + 2.4 + 2.0 + 1.3 + 3.0 + 1.5} = .0833 \text{ thousand hours}$$

or 83.3 hours.

15.31 There are $r = 4$ failures and total time on test is

$$T_4 = 3582 + 8482 + 8921 + 16303 + 296(16303) = 4,862,976$$

Since $\chi^2_{.05} = 15.507$ for $2r = 8$ degrees of freedom, the 95 percent lower confidence bound is

$$\frac{2T_4}{\chi^2_{.05}} = \frac{2(4,862,976)}{15.507} = 627,197.5$$

15.33 The probability the circuit will perform at least 100 hours is 1 minus the probability that it fails before 100 hours. Since

$$1 - F(100) = 1 - (1 - e^{-\alpha(100)^{\beta}})$$

$\hat{\alpha} = .0000909$ and $\hat{\beta} = 1.3665$, we estimate this probability by

$$e^{-\hat{\alpha}(100)^{\hat{\beta}}} = e^{-(.0000909)(100)^{1.3665}} = .9520.$$

15.35 (a) We are given that X has

$$F(x) = 1 - e^{-.01x} \quad \textit{and} \quad f(x) = .01e^{-.01x}$$

and Y has distribution

$$G(y) = 1 - e^{-.005y} \quad \textit{and} \quad g(y) = .005e^{-.005y}.$$

Consequently,

$$\begin{aligned} R &= P[Y > X] = \int_{-\infty}^{\infty} F(y)g(y)\,dy \\ &= \int_{-\infty}^{\infty} [1 - e^{-.01y}].005e^{-.005y}\,dy \\ &= \int_{-\infty}^{\infty} .005e^{-.005y}\,dy - \int_{-\infty}^{\infty} (.005)e^{-(.005 + .01)y}\,dy \\ &= 1 - \frac{.005}{.015} = .6667. \end{aligned}$$

(b) We are given that X has

$$F(x) = 1 - e^{-.005x} \quad \textit{and} \quad f(x) = .005e^{-.005x}$$

and Y has distribution

$$G(y) = 1 - e^{-.005y} \quad \textit{and} \quad g(y) = .005e^{-.005y}.$$

Consequently,

$$\begin{aligned} R &= P[Y > X] = \int_{-\infty}^{\infty} F(y)g(y)\,dy \\ &= \int_{-\infty}^{\infty} [1 - e^{-.005y}](.005)e^{-.005y}\,dy \\ &= \int_{-\infty}^{\infty} (.005)e^{-.005y}\,dy - \int_{-\infty}^{\infty} .005e^{-(.005 + .005)y}\,dy \\ &= 1 - \frac{.005}{.010} = .50 \end{aligned}$$

(c) Since $\ln X$ and $\ln Y$ are independent and each has a normal distribution, $\ln Y - \ln X$ has a normal distribution with mean $\mu_y - \mu_x = 80 - 60 = 20$ and variance $\sigma_x^2 + \sigma_y^2 = 5^2 + 5^2 = 50$. To avoid numerical integration, we note that

$$R = P[Y > X] = P[\ \ln Y > \ln X\ ] = P[\ \ln Y - \ln X > 0\ ]$$

$$= 1 - F\left(\frac{-20}{\sqrt{50}}\right) = 1 - F(-2.83)$$

where $1 - F(-2.83) = F(2.83) = .9977$ is obtained from the standard normal distribution.